Can you tell me where you are?

This book is dedicated to:

The extraordinary staff in Frimley Park intensive care who saved my life

The lady in the bed opposite me in intensive care who sadly passed away. The constant in all my memories, both real and imagined. Two people in the same situation but only one survived. I will never know why that person was me but I am sharing my story on behalf of both of us. I will never forget this lady who I never met but who I was physically close to in those intimate last weeks of life.

My family and friends who supported me throughout and continue to do so with patience, love and always humour

Can you tell me where you are?

JEN HIGGS

Crumps Barn Studio

Crumps Barn Studio
Syde, Cheltenham GL53 9PN
www.crumpsbarnstudio.co.uk

Cover design by Lorna Gray
Photographs © the author
Text by Steph Higgs reproduced with permission of the author
Extracts from ICU nurses' patient diary reproduced with the kind permission of Frimley Health NHS Foundation Trust

Typeset in Adobe Garamond Pro

All our books are printed on responsibly sourced paper from a mixture of managed woodlands and recycled material. Printed in the UK by CMP, Poole.

ISBN 978-1-915067-69-2

A NOTE FROM THE AUTHOR

Whilst this is an account of true events, some names have been changed for the sake of their privacy

FOREWORD

In 2021, I unfortunately contracted COVID-19. Towards what we thought was the latter part of the pandemic, I became extremely unwell and spent weeks in a coma on life support.

This is my story.

Within my story you will also find extracts from my mother's perspective. She lived through the same terrifying ordeal but, as my next of kin, her memories and experiences are very different to mine. From taking medical professionals' calls to keeping family and friends updated, she remained strong – even though she probably didn't really feel that way. She is a truly amazing person even though she doesn't always realise it.

And then there is the harsher reality of life in ICU with extracts from the diary kept for me by the staff who cared for me directly. These wonderful, incredible people not only saved my life, they willed me to come round. They took time to talk to me and listen to music with me even when I wasn't able to respond or when I might have been unconsciously uncooperative. They also communicated with my family with empathy and understanding when they were unable to see me for themselves.

It's taken us all a while to process the things that happened and the lasting effects following my infection. I suffered with PTSD after my coma, as did my family, and writing things down has undoubtably helped us to understand the things that happened as well as accept the changes that it brought about.

Our lives changed forever on the day of that positive test and none of us will ever forget it.

Before Covid, with Poppy and Ivy

1

31 OCTOBER 2021 – BEGINNING OF THE UNKNOWN

It was a Sunday and I was in a choir rehearsal when I got a text saying that, after being Covid free for eighteen months we had our first positive case in the nursery school where I work. It was a baby – reportedly the lowest risk age group for infection.

I probably did an eye roll and went outside to make some phone calls. I was totally calm and under control. I rarely even caught colds so I was only really worried about how the nursery parents were going to feel – particularly with some of them having health conditions that put them at a higher risk of complications. I suffer with Systemic Lupus Erythematosis (or SLE) which is an auto-immune condition and I knew the immunosuppressant medications and steroids I take for that put me at risk but didn't really consider myself to be in that category. A 'shielding' group had been announced and I hadn't been automatically put into it when Covid first hit because I didn't start the drugs used to suppress my immune system and treat my Lupus until after the list was generated. The people on the list had to stay at home because they were high risk. I had asked if I should also be careful given my new medication but had been told in no uncertain terms by the nurse in my GP practice that 'the system' hadn't flagged me as vulnerable and a lot of people take the medication that I was taking – they were fine. It wasn't until much later that I was aware that I should have been in that group.

Funnily enough the same nurse saw me after I'd had Covid. I was in a wheelchair and couldn't wear a mask because I was suffering with long Covid. "That's fine," she said. "As long as you don't have Covid because I'm vulnerable" … open mouthed I mumbled that I was fine – but *she* was vulnerable? The irony!

Anyway, we had planned for that first case at work extensively so we just needed to simply implement and follow our own plan. I had been fully involved in writing this plan and I knew every single bit of it so I was pretty confident that we would be okay.

When we were writing the Covid operational plan, it was June 2020 and we were still in total lockdown – had been for two months. Covid felt very new and very frightening to a lot of people. Schools and nurseries had been closed for a few weeks and we were tentatively planning ahead ready for reopening whilst waiting for Boris to announce that we could.

We had all seen the terrible news reports, watched the daily news briefings and seen the horrible images of people gravely ill in intensive care. We'd seen the slides with the frightening statistics on how many people were infected and how many had sadly died. Mum and I sat down together with a cup of tea and a biscuit at 5pm every evening to hear what the Government and health experts had to say about the spread of infection, the number of deaths, the strict lockdown rules and the plans ahead. It was a new kind of routine. One that we never could have imagined. It was like a dream and we just couldn't believe it was happening!

I never once related those statistics to myself. *'It won't happen to me'.* I was generally quite relaxed as someone who had developed nursery immunities during my long career in childcare and never seemed to catch infections that were circulating despite having an autoimmune disease.

As a team, we were scared about reopening. We just didn't know what to expect and I think, despite the strong safeguards we put in place, we all expected to get a case fairly quickly so with no cases until October 21, it was a surprisingly long time in coming and we were very grateful for that.

I inputted into that plan but was more worried about the team and families getting Covid than about getting it myself. As well as planning to keep everyone safe, we also needed to work out how we were going to deal with the financial impact of being forcibly closed whilst still charging fees – who subsidises that? The business or the families who use it? Can the business even afford to reopen knowing that fees which had been paid may need to be reimbursed? This was not something we had been able to plan for or consider earlier and, although we had terms that covered the charging of fees, we had very little to back us up when making decisions about this particular situation – how could we have? No one saw this coming. Some parents inevitably had to take time off, shield or were furloughed, so could they bear the financial burden? Should they? There were so many difficult decisions to make and issues to balance that the actual chance of being infected myself seemed pretty insignificant.

We decided it would be sensible for staff to get tested with it being our first case and we all agreed we needed to be swift, calculated and efficient. Keep the team calm – we'll all get tested, go into work as normal and try to prevent any further spread. Then we'll let the parents know as soon as possible.

So, even though I don't work with the children and couldn't have caught it, Mum managed to get me an appointment for a test that evening. I was really surprised to get one so quickly. I diligently went along and did my duty after the choir rehearsal so that we could reassure people that everything was okay.

The lady at the test centre offered to help me take the sample but I was pretty relaxed after having done lateral flow tests at home twice a week for months so I sat and did the familiar swab of my tonsils and my nose. It didn't even make me gag or sneeze anymore, so it was easy. I handed my completed test to her and told her that it was just a precaution and I was pretty sure it would be negative as I felt fine but it was best to be safe. I was completely confident, cheerful and not a bit concerned.

The next morning I got up and went to work as usual. No children had arrived so we were chatting easily whilst getting ready to open and I felt fine. What are we going to do? Who do we need to tell? Are all the staff ok? I recorded negative lateral flow tests on the spreadsheet as we had done twice a week for months before – most staff had also gone for PCR tests. All had come to work. Even though there had been a positive case and people were genuinely scared, they still all showed up.

Then my phone rang and it was Mum. I'd given her number to the test centre for some reason. "Jen, you need to come home. Your test was positive! You've got Covid!"

I really thought she was joking, but she was insistent. I immediately grabbed my bag and left – quickly telling the others why I was rushing out. I was furious that I had to leave – I had all my work meticulously planned out but had no choice!

"I'll do some work from home instead … I'll be back in two weeks!" I was lucky that there were some things I could do from home if needed.

On the way to the car I bumped into my friend and the nursery owner Maria. She was coming in early so that she could support the team after the first case. I stopped to tell her I'd tested positive at which she took a giant step backwards and sent me on my way telling me not to worry. I felt like 'the infected one'. I suppose I was exactly that! I burst into tears as soon as I got into the car and cried all the way home. Two whole weeks at home – and I had so much I needed to do!

When I got home I called the people I'd had contact with the day before. I was sobbing by this point because I was worried that I had infected all of them. I'd shared a car with my lovely friend Lynne who had a vulnerable family member and I was horrified by the thought that she may have caught it from me – even though I hadn't known there had been a case at work or that I was actually infected when I got in the car with her. And I had sat in a choir rehearsal without a mask around lots of people. I thought they'd be furious but everyone I spoke to was very calm and kind.

I started to feel a bit unwell that evening. The next day I woke up feeling worse but I was determined to get some work done so I sat in my pyjamas to do it. I was frustrated because I couldn't focus on it at all. I thought it was just because I was churned up about being at home so I tried to calm myself down. I knew there was plenty of work I could do from home and I had already been assigned lots of tasks so I wouldn't be bored and would still be useful. Eventually though, I had to give in and stop working because I really wasn't well.

This is where things start to become a little hazy in my memory.

STEPH

Life after Covid – what a change we have gone through! Who would have thought such a thing would happen, and that so many would be affected. We saw it unfold on the news, not ever believing that it would or could happen here. *That type of thing doesn't affect us, it's far away in China. It might spread a bit but we are safe,* we told ourselves. But it did come and it came fast! We watched the news every day and gradually realised it was quite worrying – but still not having any idea of what was to come. *We've had scares like that before,* we said.

My daughter Jen was vulnerable as she has Lupus (SLE) and Idiopathic Intracranial Hypertension (IIH) and I am an 'old senior citizen' so also considered at risk but we carried on working at the day nursery, Jen the Office Manager and me a part time receptionist. We were not told to shield when shielding first came in even though Jen is on immunosuppressants.

It did however get rather more scary when all schools and nurseries had to close their doors and Lockdown began in March 2020. I remember driving home from work listening to the news and wondering what on earth was going to happen. The nursery owner sent me home early as she wanted to say goodbye to the children and parents herself. No one knew when we would be able to reopen and how we would all survive. As I drove home the Chancellor broadcast a speech to put our minds at rest with the Furlough scheme. Like everyone else, I remember panicking and stopping off at the shop to see if I could buy masks, gloves and sanitiser!

I was so concerned about Jen that it was a relief to know that Lockdown meant she was at last able to stay at home and keep herself safe from this virus. My son, Andrew, who is asthmatic was also okay and able to work from home. The weather was great and lots of us started new hobbies. We obeyed all the rules, and there were many. We queued at the supermarkets, we sanitised and we wore masks. One way systems were put in place in food shops so that we didn't get too close to each other. There were shortages of various items which caused panic buying. Many people were shopping online for the first time which caused a shortage of time slots! Luckily we had been shopping online for a while and eventually we were given priority slots but to start with it was a total muddle.

We couldn't meet up with friends and families so we all got pretty good with technology and got used to on screen meet ups. There were daily bulletins from the Government that we sat and watched fearfully. It really did feel that we were at war but with something that couldn't be seen and that we knew nothing about. The whole world was involved and we watched very harrowing scenes on our screens.

Despite the new rules we found ourselves meeting and chatting to neighbours we hadn't really spoken to before – always from a suitable distance. We all came out on Thursday evenings and clapped for all the medical staff who were bravely soldiering on and dealing with so many awful things. The old wartime spirit seemed to come out and most people pulled together, saying *we will fight this, live through the many restrictions on our lives and when it's over we will hold street parties!*

By 2021, although Covid was still a threat, we were all vaccinated and life did get back to some sort of normality. Nursery was up and running successfully again and Jen had been back at work since they reopened in May 2020. We were still very mindful that the Covid risk was there but it didn't seem as threatening as it had been previously. Some of us still wore masks and used sanitiser, in fact for me it was hard not to, as it had been a very scary time. In my job as receptionist I remember being very strict with the teachers with sanitising pens etc. I think they thought I was a real fuss pot!

Maybe the world relaxed a bit too soon though as on Sunday

31 October the nursery had its first case of the virus. Of course there was an action plan ready and this was put into place immediately it meant that all staff had to have PCR tests on that Sunday and could only come to work Monday if negative.

So off Jen went that evening to have the PCR test – not at all worried about herself as the case was a child and she is admin so didn't work with the children. She felt fine. I tried to book a test for myself as I worked on reception but was told that it was unnecessary for me.

What a shock when the positive test came back the next morning! Jen had gone all this time without catching Covid, so how could she have got it now? On Monday morning she had gone into work as usual, absolutely sure it was okay but as she arrived the text came through with the shock result. Poor Jen rushed home very upset and disbelieving the result.

She was so sure that it was a mistake that she did two more lateral flow tests – the first showed negative but the second one showed a very faint line if we shone a light on it.

"Oh well," Jen said, "I feel okay, I've had three vaccinations so will just have to work from home for a week or two." But she was rather frustrated and upset about having to do that.

However, the next day the test showed positive immediately so we knew it was definite. And I then had to go and get tested which was not straightforward this time. There was quite a drive to get to my appointment and I was not allowed out of the car to take the test – all very stressful I'm afraid to say, but I had been in direct contact.

Now of course as I write this the Covid scare seems to be forgotten, the camaraderie that we felt in 2020 seems to have disappeared and sadly the world appears to be at war with itself but at the time Covid was the only thing that mattered. I believe that it's still a threat though. New strains are coming out all the time. We should remember that we are vulnerable and keep in mind just how quickly it can take hold. We are only just starting to see all the repercussions from it and all of the many health problems that are following in its wake.

Before Covid, with Mum (Steph)

2

DAY 7 – IMAGINING THE UNIMAGINABLE

I don't remember much after testing positive until the first ambulance was called on day 7. I'd gone to the bathroom in the middle of the night and felt really unwell. I blacked out but landed on the toilet. Everything was spinning and I didn't know what to do to make it stop! I'd been quite unwell before but had never passed out cold. As I came round, my dog Ivy was barking and the next thing I knew, mum was there. Ivy had gone to the bedroom where she was sleeping, woken her up and brought her to the bathroom. Mum managed to push me away from the door and called an ambulance. They were busy and it took an hour and a half but eventually they came.

I walked down the stairs to go out to the ambulance and vaguely remember getting onto a trolley outside the house. Mum wasn't allowed to come with me because of Covid regulations so she had to stay at home. It was scary having to go on my own. I knew I was pretty unwell and I didn't know how long I was going to be away from home. I don't remember arriving at or being in A&E. The level of oxygen in my blood was okay at this point but I do recall the doctor saying they'd keep me in overnight for observation anyway as a precaution.

Then there was a shift change and a new team took over. A

nurse bustled in and told me I was going home. I told her I was meant to be admitted and she said there had been a change of plan. I mentioned that I'd need to wait for Mum to come and collect me but she told me that I could wait in the waiting room – they needed the room where I had been in isolation.

I went out into the waiting room but I knew I was infected with Covid and my conscience wouldn't let me stay around so many other people for fear of passing the infection on. It was also heaving, someone was vomiting and people were laying on the floor – not to mention the fact that I was still immuno-suppressed from my lupus medication.

So struggling to breathe, in my thin pyjamas and in the freezing November cold, I went and sat outside and Mum picked me up. I could barely even walk to the car. I wonder if things would have been different had I been kept in and treated at that point – or had I gone back inside and insisted. I suppose I'll never know the answer to that question. There were no anti-virals for Covid available at that time but if I'd stayed in I could at least have been given oxygen. Given how busy the hospitals were, I suspect many patients like me were sent home when ordinarily they would have been kept in.

During the week that followed I was unwell. I don't remember it but I refused to go to hospital – they'd sent me home so they wouldn't be interested. I must be exaggerating. Drama Queen – I'd clearly seen too many Government briefings and statistics. I must be making a mountain out of a molehill. I'm pretty stubborn and would have continually refused to go to hospital in case the A&E staff thought I was making a fuss.

Being a burden and a nuisance has been a constant companion on my journey. Even in my unconscious mind I clearly felt like I was making it difficult for other people. Repeatedly apologising for being the one who always inadvertently caused problems, or made other people think badly of me, or even for simply existing. It's something, even now, which never leaves my mind and regularly occupies my thoughts. There's no reason behind it. I don't think anyone has ever said it to me but it just won't go away.

DAY 14

The next memory I have is of being in bed at home. I think Mum was taking my temperature or something. I knew she was worried so I tried to go down the stairs to stop her from calling an ambulance – we had been told to avoid calling 999 because the health service was so stretched.

"I'm fine – look, I'm up and coming downstairs!"

But I really struggled to catch my breath when I moved and had to stop after a few steps. '"Okay, I will email 111," I relented, as I was attempting the stairs – barely able to breathe. Mum called them in the end and a doctor called back really quickly. He insisted on speaking to me. He then called for another ambulance. They came quickly this time – no blue lights, just minimal fuss which I was grateful for.

Two female paramedics arrived. They asked me to walk around the living room to check how breathless I was – I know I was struggling but, in my head at least, this was all a waste of time. I'd been to A&E a week ago and they'd said I wasn't ill enough so I obviously just needed to be brave and ride it out.

I didn't think they would be able to carry me so I walked out of the house and up the drive to the ambulance. I was very breathless and extremely aware of getting too close to the paramedics for fear of infecting them. One of them was asking me questions whilst the other drove. I was sitting up in the seat in the back of the ambulance – not on the trolley or in a wheelchair. I was a passenger rather than a patient. She was asking me things and inputting my answers onto an iPad. The whole time I was trying to answer honestly but, as I was so convinced I was just going to be sent home again, I probably downplayed how rotten I felt. I think I tried to crack a few jokes along the way but she didn't laugh. Says it all about my sense of humour I suppose.

We arrived and the paramedic told me there was a long wait to go from the ambulance into A&E so I would be waiting for a while. They left me there and went inside. I don't remember sitting in there waiting to be taken in, getting out of the ambulance or going into A&E so have no idea if I blacked out or if I just don't remember.

Memories become even hazier now as I have very little recollection at all of being in A&E and had no one with me.

Breathing was becoming really difficult and I was bleeding from everywhere. A blood test confirmed that my INR (International Normalised Ratio, which indicates how long it takes blood to clot) was over 10, rather than the 2.5 we were aiming for, so my blood was way too thin and I needed vitamin K urgently to counteract the warfarin I had been taking following a pulmonary embolism earlier that year. I was having stomach pains and my stomach was quite hard. A scan showed internal bleeding so surgery was discussed. I don't remember what happened to stop this but I assume I was too unwell by now to undergo any surgery and the doctors decided that they would wait to see what happened with the bleeding rather than operating.

STEPH

As the week went on Jen seemed to develop cold like symptoms and became extremely tired. Work was becoming a struggle for her and she started to become breathless. In fact the decline was really quite fast and by Saturday 6 November she was very poorly – sleeping a lot, struggling to breathe, not eating or even drinking. It was impossible for me to keep a safe distance from her. I tried my hardest to care for her but it was frightening watching her deteriorate to the point where every breath was a struggle.

In the early hours of Sunday morning I was woken up by one of the dogs (Ivy) barking frantically. Ivy wouldn't stop so I got up – feeling very cross with her for waking me up! She ran to the bathroom door but as I pushed, it wouldn't open. I glanced in Jen's room (next door to the bathroom) still half asleep myself and I was surprised to see that she wasn't in bed. Suddenly reality hit me I found she had collapsed against the bathroom door seemingly unconscious and making a very strange and frightening rasping noise.

I forced my way in and pushed her back to a sitting position which seemed to relieve her breathing but she didn't immediately respond to me. It was all very scary – it bought back the horror of what happened when her father passed away very unexpectedly aged just fifty-four in 1998. Fully awake now and shaky but trying to keep calm, I grabbed my phone and rang 999. At that she started to come round a bit more and begged me not to call. I don't think she realised what had happened – she was worried about making a fuss and didn't want me to phone for help but I knew we needed help.

Eventually I managed to get her back to her bed. The phone

operator told me to put the dogs out of the way and to have the front door open ready for the ambulance which I did but there was a very nerve wracking hour and a half wait. I felt completely out of control. All any mother wants is to be able to comfort their child and make everything better for them but I couldn't do anything for her apart from talking to her, stroking her head and reassuring her that it wouldn't be too long before she would get some help.

The paramedics were brilliant. They checked Jen over and told me to pack her a bag as they were taking her in and they believed that she would most likely need to stay in. Her blood pressure was very low, her breathing was difficult and she had a fever. She really was very unwell.

Obviously I wasn't allowed to go with her due to Covid restrictions. What a strange feeling when they all left – a cup of tea was most definitely needed but I had no idea how I would know what was going on, where she would end up or even who to phone to check on her! It was still the early hours of Sunday so I decided to wait a while before letting my son know.

I waited all day to hear something and spent the time staring into space thinking constantly about what she was going through. I believe that I did eventually ring the hospital and it was confirmed that she was still in A&E and had been having tests. Later Jen managed to call me to let me know that she was on IV antibiotics, had a touch of pneumonia and that the doctor said they would likely keep her overnight, just as a precaution. I felt a sudden rush of relief that she was in the right place and would soon be back home feeling much better.

But I continued to worry and pace around until early evening when Jen called and asked me to pick her up! She had been discharged into the main (very crowded) A&E waiting room – even though she had been kept isolated up until then and was

Covid positive! – so she took herself outside in the freezing cold wearing just her pyjamas to wait for me.

By the time I got there she could hardly walk to the car (even though I had managed to park really close to her). Her breathing was very bad and she was really struggling. I was so worried and I wanted her to go back in but there was no way that she would and she was very upset. Looking back now I feel terrible that I didn't insist that they take her in but at the time A&E was so busy that there were huge queues around the outside of the department. The poor staff must have been having a real nightmare. But should I have insisted? I feel I should have, but at the time seeing how busy they were, thinking they must know best and seeing how upset Jen was we just drove home. I wanted to get her safely in the warm and wrap her up until she felt better.

She had been given no medication, just told to rest so that is what she did but as the week went on I watched and worried as she got worse and worse. After her experience that Sunday, Jen was adamant that she would not bother the doctors again. She is very stubborn. She was certain it would start getting better – after all they had sent her home hadn't they? But she wasn't eating – even drinking was difficult. So I kept quiet but was very worried as she got weaker and weaker, sleeping much of the time. By Sunday 14th when she woke, she could hardly breathe or speak so very much against her will I rang 111. Luckily they answered immediately and got a doctor to call back within five minutes. The doctor asked to speak to her and he promised that an ambulance would be sent to us as a priority. He said to pack her a bag again because he didn't like the sound of her breathing and he was concerned that she wasn't completely coherent and was a little confused.

The ambulance arrived in fifteen minutes this time. I couldn't believe it! There was hardly any time to get the bag packed! They assessed her and decided to take her back to the hospital. The crew were quite calm and said it was just precautionary. Here we go again … She didn't want to go but could hardly breathe or move. It was very upsetting to have to wave her off once more. This time I knew it was bad but I wasn't at all prepared for what was to come. How could I have been? Things like this don't happen to us. Do they?

Once again I had a long day of waiting and wondering, not knowing what was going on. This time she was eventually admitted to the Covid ward (no visitors allowed!) – just for 24 hours observation they said. I felt so relieved at that news and reassured myself all would be well, she was in the right place.

3

ADMITTED TO THE GIANT PEACH

I was admitted to the hospital and put onto a Covid ward. There were four of us on my section of the ward. I somehow discovered that the two ladies opposite me were mediums and I can remember telling them that Mum was a spiritualist. They knew some of the same people that she did. I don't remember them being unwell at all – actually for some reason, in my head, they reminded me of the two comedy sisters from James and the Giant Peach – in pyjamas! They obviously knew each other well – I thought how lucky they were to be on the same ward with someone they knew and not to be alone.

Soon I began to need more oxygen than the normal flow mask and was put onto a high flow CPAP (Continuous Positive Airway Pressure) machine which pushes oxygen more forcefully into the lungs. I remember being told it would be noisy and really overwhelming when it came towards my face but no one could have prepared me for what it was actually like. It was similar to the feeling I imagine from sticking your head out of a moving car window on the motorway – I mean I've never actually done that but as I imagine it! But that frightening feeling pretty much stopped immediately once it was on my face and then it was like some sort of heavy suction so each time it came off I tried my hardest to stay calm whilst it was put back on – knowing that once the horrible part was over it would help. It was so frightening.

During this time I still insisted on walking to the toilet and refused any suggestion of a commode or (God forbid) a bedpan! But I was in HDU (High Dependency Unit) now as I was requiring more and more support with breathing and was soon put on strict bedrest.

Through all this time no visitors were allowed because of Covid so I had to deal with everything by myself and then report back to my family. Covid regulations, whilst completely necessary, were cruel and punishing and being unable to have physical contact with loved ones was very isolating and lonely. When you are so very unwell and vulnerable you need family there to advocate for you, to comfort you and remind you that you are loved. But I was alone – alone when I was bleeding, struggling to breathe and seriously starting to wonder if I might die (although I don't think I ever really believed that would happen). I still thought I was just being a wimp. When you've been dismissed your whole life by doctors you start to believe that you must be a hypochondriac and that other people must be able to tolerate far more than you can. That you're less worthy of treatment because a small minority of doctors all through your life have made you believe that even though you may feel unwell, you're just not ill enough or that you are imagining it all. That other people are much worse and that you are just weak. You should be thankful that things aren't worse right? Except that doesn't help when you do actually feel unwell. Maybe this is part of the reason why my dreams and thoughts all revolved around me being a burden.

Sadly, this issue is an all too familiar story with lupus sufferers. The illness imitates so many others and symptoms are so systemic that it relies on someone putting them all together and seeing the bigger picture. Unfortunately, it tends to be a case of treating these individual symptoms as they arise and often it falls to the patient researching themselves when they start to get desperate. This of course has its own pitfalls as if you allow yourself to focus on one specific diagnosis that you think fits your symptoms, it immediately puts you in danger of discounting anything that you then think doesn't fit with your 'diagnosis'. But those symptoms you discount could be the key to finding out what is really going on.

Ultimately, diagnosis often takes years if it ever happens at all and doctors are so busy that they are only really able to treat what they can see and manage symptoms. There needs to be one person with medical knowledge looking at that patient's symptoms and referrals then putting the puzzle together. It takes up a considerable amount of time and would need specialist knowledge. Who has time for that?

STEPH

On the Covid ward Jen was able to keep in touch with text messages and phone calls but she was very weak and breathless. It turned out to be more than a 24 hour stay as she got weaker and weaker. I was not able to visit but there were things that she needed so I would make the journey, park up (not always that easy) and pop anything she needed into the PALS office on a fairly regular basis. Life got really busy for me with seemingly endless phone calls, messages, emails, messenger and WhatsApp

– so many kind people concerned about Jen but replying to them all seemed exhausting and very stressful at the time.

For poor Jen, each day seemed to bring her another problem. She was already on quite a high level of intervention as her oxygen levels had become very low. She had pneumonia in both lungs, a high fever and a very painful and swollen tummy. Her tummy swelling turned out to be internal bleeding most likely caused by her blood thinners and on 17th November she needed a blood transfusion as she was dangerously anaemic. Even then she was classic Jen and was asking me to take in her walking stick so that she could get herself to the toilet! As her mum, and having watched her deterioration, I really wanted to do as she asked but in my heart I knew she was clutching at straws. It felt so hard and frustrating not to be there for her knowing how distressed she was.

But complete bed rest was the rule now which was hard for her to understand. I began to get some very sad phone calls from her. She really struggled with how sick she was becoming and how she was now confined to her bed. She must have been so frightened and I couldn't do anything to make it easier on her.

Calling me was difficult now as she could hardly speak. She was also not always making sense and was a bit delirious some of the time. It seemed so cruel that none of her family were allowed to see her and it was so difficult taking those calls and hearing her struggling and crying knowing that I couldn't do the very thing I wanted to do which was to be there – even if it was only to hold her hand. I remember one day after I had dropped something into the hospital Jen rang just as I started driving out of the car park. She was in tears and could barely speak. She was so breathless she needed the mask on all the time – but she of course tried to take it off. I really struggled to understand her. I could hear her getting so frustrated and it was incredibly distressing!

After that call I remember driving home in tears myself – I couldn't believe what was happening. I was helpless when she needed me most and totally reliant on the amazing people who were caring for her. I had to drive away knowing full well where this could lead. My heart was aching. We had all seen the Covid news reports with images of people who were exceptionally sick. I hated to picture her like that.

I phoned the ward but didn't really learn anything more about her condition. The ward staff were always so very busy whenever I rang in, so I wasn't really getting many answers as to what was happening. It was very scary at the time. They of course were struggling to cope with all that they had to deal with but it was also a difficult time for the whole family and we needed reassurance that she was going to be okay which nobody could give us. I felt as if I was living in a dream and this couldn't be happening. It was so very hard not being able to be there to give support and talk to the doctors caring for her. So many people told me they were praying for her which was really kind and I honestly believe that in the end their prayers did help.

Many friends and family were around for me, but no one was there for her – she was alone!

Silly as it sounds I too felt so alone (which as I said before, I wasn't). My son, his partner, my brothers and friends were there for me – if not with me they were on the end of the phone. But in a strange way, I think I felt as if it was something I had to deal with and I didn't really open up about my feelings – bottling it all up inside and denying that I needed help. When my son offered to stay with me I turned him away not wanting to be a burden. Should I have let him? Maybe I should have as he must have been going through trauma as well.

At that time I was imagining how scared Jen must be feeling and I became unable to concentrate on anything – I couldn't eat or relax as I anxiously waited for the phone calls from the hospital. I found myself wasting time just sitting staring into space which is not like me at all, then feeling guilty as I wasn't getting around to housework etc! So many people looked out for me – walking the dogs, bringing meals etc. I was living in a dream – a nightmare. It was a fear I had right from when Covid first started to hit the UK and I couldn't believe it was happening to us.

Jen soon progressed to a CPAP machine so then it was very difficult for her to talk and when she tried she was very confused and difficult to understand. At one point she said, "this is horrific! They have put a huge mask on my face! I feel like this is the end of my life."

It was so hard for a mother to hear that and so sad for a daughter not to have the comfort of her mum when she obviously so desperately needed it.

Overnight on 21st November Jen was taken into HDU as she was really struggling and needed very careful monitoring. She was also very dehydrated and was now sleeping much of the time. Her calls and texts to me had become a lot less frequent and lot more difficult. When she tried to speak she was begging me to bring drinks to her. I felt in such a quandary! How could I ignore her pleas? I knew that she probably didn't really need more drinks taking in by that stage and I still couldn't see her. But I did go and get drinks which I handed into the PALS office only to be told that we were now not allowed to send in food and drink to the patients. They took pity on me though and said they would send them up to the ward just this once. I discovered later that those drinks never did get used but she would have been unable to drink them anyway by this stage.

At 2am on 23rd November I was fast asleep when my home phone rang by my bed. In my sleepy head I thought it was my alarm and tried to switch it off. By the time I realised it was actually a call I had missed it! Oh my goodness! I checked the caller number and what a shock it was to see it was the hospital.

Shaking all over I tried to ring back but at the same time my mobile rang and there it was – that dreaded number! I tried to take in what the doctor was saying. Apparently they were moving Jen to ICU as she was going downhill and they needed her to be where there could be a higher level of support.

"Am I going to lose her?" I asked quietly. The doctor didn't answer for a short time then said there was possibly a 50:50 chance that she would get through this and that ICU would be the best place for her to be. The doctor asked if I could come to see her as soon as I possibly could before they move her.

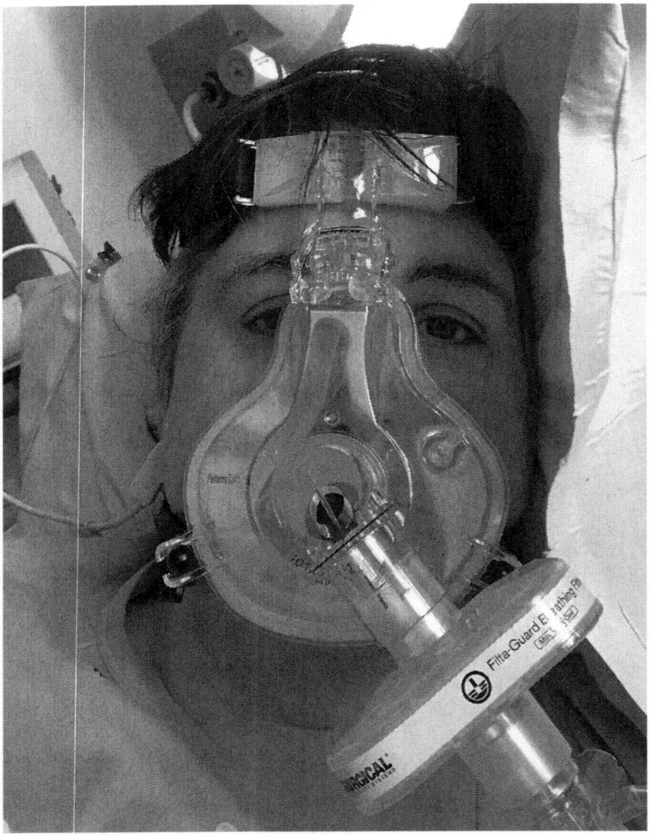

On the CPAP machine

4

DAY 21 – SAYING GOODBYE

Mum was called to the hospital at 2am because I was being moved from the High Dependency Unit (HDU), where I was on the CPAP machine, to Intensive Care (ICU) – in case I needed ventilation.

When she arrived I was being wheeled along the corridor to ICU. I was in a tent because I needed the CPAP machine which sprays out droplets and I was still infected with Covid. Mum had to wear full PPE (and this means head to toe covering). I don't know what my oxygen saturations were at this point but later on a doctor told mum I was doing well with sats of 35% so it must have been pretty dire (they were normally 98% to 100% for me)

Mum and I spent a little time together but I was much less able to communicate with her due to my breathing. I was surrounded by equipment and machines. I was worried about having to have a catheter but there was no choice. Apparently, I asked her to tell my brother Andrew that I love him and to tell them to replace me at work. I don't have any memory of this.

DAY 22

As breathing became more difficult and I became more and more exhausted, the decision was taken that I needed to be ventilated. I don't remember, but have been told that I was very resistant to this however as I got weaker it became less of an option and more of a necessity until I was so sick that I was asking for ventilation.

I was put onto a ventilator and into a coma. I had pneumonia and soon after, I unfortunately also developed sepsis. My condition was dire and survival was not a given. No one knew whether I would wake up and, if I did, whether I would still be me. What sort of life was ahead of me?

STEPH

In a dream and shaking I threw on some clothes and let the dogs out quickly as I didn't know how long I'd be. They must have wondered what an earth was going on! It's a half hour drive to the hospital – should I ring Andrew to let him know? Would he want to see her too? Would he even get there in time? Should I worry him with this in the middle of the night? Would he even have his phone on? I didn't have much time to think but decided not to make that call as it would take him over an hour to get there anyway and by then it could have been too late for us to see her. In hindsight I also think he may have found seeing her a bit overwhelming because she was so unwell.

It was a very smooth quick drive for me – absolutely no one else seemed to be on the roads. On getting to the hospital the car park was very eery and empty, the doctor had said there would be someone on reception to point me in the right direction. Thank goodness I thought, as I really hadn't taken in the directions that she gave me. But no, there wasn't a soul around.

I tried to remember what the doctor had said and followed my nose. I remembered that she had said I should be going towards the private suite so I followed the signs. They led me to a lift. Well, I'm scared of lifts having had a bad experience once. I just stood there in tears pressing the button. I needed to get to Jen and I knew I was going to have to be brave.

Suddenly I heard voices and two lovely nurses came along. What a relief! They were so kind and when I explained why I was there they took me straight to the High Dependency Unit where we found that Jen was just in the process of being moved. Was I too late? The doctor was very kind saying that I could follow them to Intensive Care and she walked with me as we followed the bed (I'm a bit slow!). There were three medical staff walking with Jen who was in the bed with a mask over her face. It looked so frightening and so very serious. They had what appeared to be a tent over her because she was still testing positive. Halfway there we had to stop while they changed the oxygen. If it hadn't already hit home how unwell she was it certainly did now.

Once we got there Jen was whisked away and I was taken to a relative's room and asked to wait while they sorted her out and made her comfortable. She needed to be hooked up to monitoring machines and the oxygen supply had to be changed over. It seemed like hours (I guess it wasn't really) but eventually I was taken along what at the time seemed like a very long corridor. Having since revisited there, it really wasn't that far but at the time it felt like miles.

I had to be dressed up almost like a spaceman and was taken into the room where poor Jen was still on the CPAP machine, hooked up to monitors with a nurse keeping an eye on things. Jen was not quite with it but she did try hard to talk to me. It was difficult but I was so glad that I had been allowed in. I was able to hold her hand and tried to reassure her but she was frightened.

She said, "Am I going to die?"

"Of course not," I said, with my tummy churning all the time, "this is only in case you need a bit more help with breathing."

Clearly she didn't believe me as her next words were, "Please tell Andrew and Phil [her brother's partner] I love them and tell Maria [her boss] that she will need to replace me, I don't think I'll be going back to work."

She told me that she loved me but by then she was so exhausted, it was a real struggle for her to speak. Sadly our time together was short which was hard but I knew by then that I was in the way. The nurses had their job to do and she needed rest. Sad and difficult as that visit was, I am so very grateful that I was allowed to be with her before she was sedated and ventilated. I didn't know when or if I would see her again. Was I really about to lose my daughter?

I had to partly de-robe before I went out of her door and re-mask up to walk out of the ward then I was taken back to the relatives room. The nurse had a chat with me there and gave an idea of life for Jen in Intensive Care. She said that, although Jen had been adamant that she would not go on a ventilator, she had been moved just in case that was needed. She also needed closer monitoring and would have a nurse assigned just to her 24 hours a day. She assured me that now Jen was in Intensive Care I would hear from a doctor daily to keep me informed.

Walking away to the car park in the dark, the seemingly empty hospital felt very strange. The roads were still empty and quiet – I think I drove home on autopilot.

I got home at about 5:30am to two very mystified dogs and wondered whether I should ring Andrew or wait until I knew he would be up. I laid on the bed feeling shattered and just waited.

When I called he came straight over to be with me. He

took me for breakfast and we chatted. I think we both felt in a dream – this couldn't be happening could it? After all the health problems Jen had been dealing with all these years, we couldn't believe she had been dealt such a difficult hand – and after two years of restrictions we had all thought Covid was not such a threat anymore.

On the 24th we were told that Jen was very seriously ill. Her blood pressure was dangerously low and her chances of surviving were dwindling. She had been put on the ventilator overnight. That afternoon Jen's very good friends Lynsey, Derek and their sons Mark and James (Jen's Godchildren) came over to see me. This was a very emotional visit and finally it was impossible to hold back the tears as I told Lynsey the news.

The days that followed were very worrying. Every time the phone rang was nerve racking – was it that dreaded number with more bad news or another well-wisher? Most often I would wait to hear from a doctor but did sometimes end up calling myself and talking to the nurse looking after Jen that day. It was so difficult knowing how poorly she was and not being able to be with her. I couldn't help with her physical care of course but just to be there holding her hand would have been enough. Just to see what was happening for myself. To be there willing her to fight.

I fully understand that all the hospital staff were rushed off their feet, but most times when I called, the nurse on duty couldn't tell me too much and I sometimes got confusing or conflicting information. I found myself making notes of all the conversations with the hospital staff and then decided to start a diary – that was a very good decision. I would strongly advise anyone in a similar situation to do the same thing.

The diary helped me at the time and once Jen felt ready to, it was very good for her to read while she worked to come to terms with what had happened to her. She doesn't remember very much from that time, her memories are muddled and intertwined with dreams and delirium – all very scary for her as the memories pop back into her thoughts. It was quite a while after she regained consciousness before she could bring herself to look at the diary but I hope it helped her in some way. For me, I needed to see it written down. I needed to be able to revisit what I had been told and to relay it to friends and family. The only way I could process it was slowly. Digesting every word and being able to look back over my notes as time went on. It gave me a purpose. I couldn't physically do anything for Jen and I needed just to do something. Documenting it made sense.

During the next weeks in ICU she had to be regularly turned from laying on her back to her front and then back again a few hours later as she was so unwell. I don't know how they did this with all the equipment around her but I'm sure they had become quite used to doing it through the pandemic for patients in acute respiratory distress because it helped to expand their lungs and improve oxygenation. But by 29th November her oxygen levels had risen to 40% and it was decided to try and lower the sedation. It was hoped that she could be safely and gradually woken up.

On 1st Dec we got the great news she had been taken off the ventilator and was back on the CPAP machine. The nurse put the phone to her ear and I spoke to her. She couldn't respond but the nurse was sure she could hear me. How confusing that must have been for Jen! I had to be positive and encouraging, despite still being absolutely terrified. It was extremely difficult having a one sided conversation with no real idea of whether she could hear and understand what I was saying. I so badly wished I could see her.

On 2nd December we were told more great news. Jen had been lifted out of bed to be sat on a chair. The doctor said she was very weak, couldn't speak but it was starting to look very positive. She had turned the corner amazingly! Suddenly there was hope and good news to share. I felt a glimmer of hope and started to wonder whether she could actually beat this.

5

TEMPORARY LIBERATION

As time moved on I started to improve a little and things were looking more positive so I was taken off the ventilator for short periods of time.

All I can really remember is a speech and language therapist coming to visit me to look at swallowing. It must have been early on – after the first ventilation but prior to the tracheostomy because she said I could have a yoghurt and it was honestly the most delicious yoghurt I'd ever eaten! I was hungry and I wanted more! I had been unable to eat for a while.

Unfortunately, the pleasure of eating the yoghurt was short lived. I was soon too unwell again and needed to be put back on the ventilator. I had managed 24 hours of breathing on my own but it was too much. I got very much sicker.

STEPH

Unfortunately that great news was very short lived. On 3rd December, just as I was off to go and read to the children in nursery (I had stopped working but was popping in weekly to see the children) I got the call. Jen was back on the ventilator.

She had taken a turn for the worse and the doctor was very concerned. Her heart was struggling, she had serious lung problems and her blood pressure was very low. I carried on to nursery but dissolved into tears as soon as I arrived. Everyone was so kind and caring. I couldn't read to the children that day and haven't been back to read since, I'm sad to say.

On 4th December I got a call to say that Jen was now on dialysis as her kidneys were failing. She still had serious lung problems and her heart was struggling. The doctor was very concerned. That was such hard news to take in but just at that moment my brother Steve and sister in law Julie knocked at the door. They had bought food and sat me down with a cuppa and offered the comfort that was so very much needed at that moment.

I still find it strange that Lynsey and her family had arrived at just the right time before and now here was Steve and Julie also with me at exactly the right moment. Was that divine intervention looking after me? I was so incredibly grateful for all of them at the time.

On 6th December we were still finding things very difficult. We were getting conflicting news and speaking to different nurses each time we rang so Andrew and I requested to actually see a doctor so that we could ask the many questions that we had. Sadly even then we were not allowed in to have that face to face chat so a phone appointment was booked.

Andrew came over to me and we sat for what felt like a very long time waiting for the call, notebooks at the ready. Eventually a very kind and understanding doctor did ring and answered all our questions. Yes it was very serious. She had internal bleeding, low blood pressure, multiple organ failure and was on very strong antibiotics. He assured us that she was heavily sedated and would be unaware of everything that was going on and that they were doing all that they could. He said that after this there will be a long road to recovery with long term health problems on top of the lupus that she had already been struggling with.

We really appreciated the time he took to explain things but it was utterly heartbreaking. What would her life be like if she did recover?

Now I was finding myself completely unable to concentrate on anything – even basic day to day chores were not getting done. I was just sitting around aimlessly much of the time. But I did busy myself putting together CDs of Jens favourite music to take in for the nurses to play to her. I also asked family and friends to record messages to be played in the hope of her being able to hear them.

The day after I took the CDs in I got the news that now her eyes were swollen and infected. The doctor thought her corneas may have been scratched by the need to turn her several times.

Feeling upset I questioned why then did they have to turn her? The answer was "to save her life!"

They had to tape her eyes closed after that.

On 18th December the doctor rang. He was concerned, saying she was very seriously ill. She had what looked like a fungal infection on the inside of her right elbow. They did tests and sepsis had been confirmed. Even stronger antibiotics were needed.

In this conversation the doctor also said that they really need to try to lower the sedation and get her off the ventilator within the next two weeks. They hoped to be able to perform a tracheostomy to make waking up easier and lower the sedation to see if she would come round.

He showed concern for her quality of life if ventilation was to go on for too much longer.

I remember going to the annual craft fair (an event that we both loved) but when I got through the door I found it was heaving with people, no masks were being worn and I just wanted to shout at everyone! I just couldn't bring myself to stay in that environment so, much to the annoyance of the door staff, I went straight back out through the 'in' door. I was told to go back and follow the directions to the way out but (unusually for me) I stood my ground and said no way would I go back in there and I complained that there were no restrictions on numbers and no masks worn. "Well the rules have changed now," I was told. Well I was so upset I found myself telling them about Jen – they let me out then!

By the 20th, amazingly Jen had taken some breaths for herself but the sepsis had worsened and a different bacteria had developed in her lungs. It seemed that the infection on the inside of her elbow was the source of the sepsis.

On 23rd December I had a very difficult call where the doctor asked whether we would put a DNR in place as she would be too weak to withstand resuscitation. Her lungs and kidneys were badly damaged by the Covid. She was fully off sedation now but showing no signs of waking up. They were going to do a brain scan and promised regular updates. I did give permission but only after speaking to my son. We both agreed that is what she would want if it came to that.

We had spoken fleetingly about things like this in the past and she had always seemed to believe that if things were going to worsen with treatment and not improve then treatment should be withdrawn. We had never had that actual conversation though so we could have easily made the wrong decision. As it turned out it was the right decision and she completely agreed.

This period of severe illness has taught us that life is fragile and can be switched off at any moment. So talking about these things before they happen is absolutely necessary. Not an easy conversation granted but easier than making this decision if I had no idea of her wishes.

6

23 DECEMBER 2021 – DETERIORATION AND FADING HOPE

A definite low point as mum and Andrew were asked to put a DNR in place.

Things were beginning to look bleak as my body was now shutting down. The doctors made it clear that I was unlikely to survive any intervention should my heart stop. If I did survive, I would most likely be severely disabled. My kidneys had failed and I needed dialysis and my heart then also began to fail. My lungs appeared to be full of a sticky substance and the doctors weren't altogether sure what it was. My chest x-ray from this time is pretty terrifying – even to my untrained eye. They were very clear that, should I survive, I would need a lot of care.

Whilst I was in the coma I developed a large ulcer inside my elbow which started with a cannula and it was thought that this could have been the root cause of the sepsis. It began life as a blister then grew and grew into a 10cm wide and 1cm deep crater. After reviews by dermatology and swabs and samples it was eventually diagnosed as *Pyoderma Gangrenosum*, another auto-immune problem. My body was eating itself and the ulcer grew at a rapid rate whilst I was given intravenous (IV) antibiotics and steroids. That ulcer took eight months to heal and required daily medications and dressings.

I developed chemosis in both my eyes and it was thought that they may have been scratched whilst in intensive care. ICU patients with Covid are proned (turned onto their stomach) during treatment in order to help breathing. The doctors told mum that this was done to save the life of the patient at critical points. I was proned numerous times whilst I was ventilated. Given the amount of equipment involved in caring for an ICU patient, turning them must be quite difficult and this could have been when my eyes were inadvertently scratched. The ophthalmologists advised that my eyes should be taped closed to protect them and they reviewed me every day or so to see how they were healing – waiting to decide if anything else needed to be done. They warned mum that I might have permanent damage to my vision.

After a few more weeks on life support, I started showing some signs of improvement and a tracheostomy was put in place so that sedation could be slowly reduced to assess my condition whilst ventilation continued.

Once sedation was stopped however, I was still not responding and I remained that way for about a week despite medication to keep me asleep being totally withdrawn. A scan of my brain revealed that I had suffered a stroke but it is unclear exactly when this happened.

More and more is being discovered about Covid and it's far reaching effects and, I'm by no means an expert but bleeding or clots seem to be a fairly common occurrence and so far this apparently isn't well understood. Unfortunately I experienced both the bleeding and the clotting to it's extreme during my infection and indeed in the months beyond.

STEPH

Goodness how my heart lifted when on 24th December I had a call from a physio who was asking all about Jen's life, hobbies etc, as all this info would help with rehabilitation. She also said that Jen had been lifted onto a recliner chair. She suggested that we take in photos so that they could talk to Jen about familiar things. I imagined her sat up, listening to them talking but sadly I think the reality was probably very different to what I had hoped for at that moment!

The next call was from a Rheumatologist asking about her lupus so that he was better able to help with her recovery. Amazing, I thought, and I told everyone the good news! Things were moving forward as they both seemed so positive. I too felt suddenly very positive. She might actually be able to beat this! I so wanted that to be the case.

Unfortunately that evening the doctor rang me again to let me know that indeed Jen was still very, very poorly and in danger. The DNR was still absolutely necessary. Jen still had kidney failure and was now relying on dialysis, her heart was struggling, the sepsis was not improving and was a big concern. I couldn't believe what I was hearing after the calls I'd had that morning. I still don't fully understand how I was given such mixed messages. It felt like utter devastation all over again. Not being able to see her for myself had massive pitfalls because I had no real idea what was going on. Was she taking one step forward and two back or was it the other way around?! I had no way of knowing for sure and had no choice but to rely on what I was told.

Jen was off sedation but the doctor explained that she wasn't responding despite being sat up in the chair. Apparently the scan they did showed that she had suffered a stroke but couldn't really explain why she wasn't coming round. The doctor couldn't say when that had happened but it would most likely have been while she was sedated and on the ventilator. So she was sat up in the chair, not moving, staring into space and totally unresponsive. Not what I had hoped for at all. I couldn't bear to think of her like that. Could she see? Was she stuck in there unable to move or communicate?

This news was so hard to take in and we were all very upset. Tomorrow was Christmas Day, Jen was still in a world of her own, and the doctors didn't know when or if she would wake up. If she did, they didn't know if she would still be the Jen we all know and love. Nor did anyone know what the effect of the stroke would be. It was the reaction to Covid that we all feared but all believed that we would be fine. We all tend to think things like this don't happen to us.

We decided that Christmas was cancelled for us but Andrew, Phil and I would just be quietly together on the big day. No presents we said!

I insisted on cooking a meal but it was not the normal big Christmas dinner, more a shop bought readymade meal from the freezer. Christmas is always a lovely family event that we all enjoy. Jen loves all that Christmas brings. She puts so much love and effort into the gifts she gives and the beautiful food that she cooks (taught by me of course). For some time now she has been in charge of making all the special Christmas foods. She would be so upset to be missing it.

So for us it was a very odd, quiet, sober sort of day, wondering just how she was ever going to be back with us. I couldn't imagine what life was like for her at this point. She wouldn't have had Christmas dinner or presents. Actually I found out afterwards that she did have a present from the nurses which was lovely. A pair of pink fluffy socks! She was very confused about their origin when she did eventually come round.

On Boxing Day evening, Andrew and I went to the hospital with more pictures and messages. The physio had suggested that once she began to come round they would show and talk to her about them to try and help her recovery. We stood outside the ward as usual waiting to see someone and eventually we were so pleased when the nurse looking after Jen that day came and chatted to us. It was lovely to see the person who was caring for and knew Jen. She did give us a little ray of hope when she said, "I thought I might have seen a very slight response from Jen today when I spoke to her."

She also confirmed that they had been playing the music I had sent in. I know the messages were also played to Jen but she has no recollection of them. She is still unable to listen to them even though she is grateful that they were played to her. Whilst lovely messages, they continue to bring scary thoughts to the surface.

I was told that she'd had another blood transfusion and another line had to be put in for the dialysis. On the 28th they lifted her into a special recliner chair where she stayed for forty-five minutes, seemingly calm with eyes open but still not responding. Could she know what was going on and not be able to respond? I just couldn't get out of my head what a scary prospect that was and I couldn't help worrying about just how frightened she would be – also knowing that I should be there talking to her! I just kept wondering if she would have responded to my familiar voice. If the touch of my hand could have brought her back faster. I'll never know but being away from her and knowing that I couldn't have made a difference was agony.

Finally a few days later a doctor called me. Things hadn't really changed too much but he had seen her move her head very slightly and she seemed to be taking some breaths herself. He said physio would spend more time with her now and the blood transfusion was quite normal when someone has been on long term dialysis.

On 30th December we received amazing news! Jen had been off the ventilator for fifteen minutes and some eye contact was noted. She even managed a small arm movement. The stroke was on the right side of her brain so that means she may have some damage to her left side. The doctor was still very concerned about the infection on her arm as it appeared to be turning into a severe ulcer. But at last good things seemed to be happening.

7

OUT OF THE DARKNESS AND INTO IMAGINATION

Suddenly, after a week of being unresponsive, one of the nurses thought she had seen my eyes move. It wasn't like the movies or a soap opera where the patient opens their eyes, sits up and says thank goodness that's over. No, it was much darker, lonelier, scarier and totally crazy at times.

It was completely dark and, as if from the Gods, someone was saying, "Jennifer, you've had Covid and been very unwell. You're in intensive care." It was the first time I had any idea of what was going on and it must have been really early on in the waking up process. All I could see was black. I couldn't quite comprehend what they were saying to me. How was this even possible? I stayed still for a while. My brain felt really sluggish. I couldn't get it to focus on anything. What do you mean intensive care? That couldn't possibly have happened to me. Could it?

I eventually opened my eyes and the lights were very bright. For some reason I couldn't move at all – I felt a bit like I had been swaddled in a tight sleeping bag with only my face showing like a Russian doll. I became aware that someone was asking me questions but she didn't seem to be speaking in a language I understood. I couldn't work out where on earth I was. I tried to speak but couldn't make any sounds. Something was very wrong with my mouth and it wouldn't move in the way I needed it to. No sound was coming out. I must be dreaming!

Gradually, the person asking me questions started to speak in a language I understood. She was trying to get me to respond but I really couldn't. I didn't know what to do. I just couldn't get my mouth and my brain to co-operate with each other. My brain was racing round and round trying to understand what had happened to me.

She went away and came back with a laminated piece of white card. It had letters on it in black ink and I was relieved that I did recognise them. She pointed to one and somehow I was able to indicate with my eye movements that it was the correct one to spell what I was trying to say. Short simple words. Slowly we spelt out words together that I would have said if I was able to speak. It took a very long time! She was standing to my right and I couldn't see anything except for her face and the laminated chart. It was very slow and laborious. She was exceptionally patient.

This feels like it must be a glimpse into what it would be like to be locked in. It's a terrifying and lonely place and it feels like being in prison. Imprisoned by my own body. In a parallel universe watching everything carrying on as normal around me but I was unable to participate in reality. Still in a state of disbelief that this had even happened to me.

For me, I was blessed that it appeared to be short lived as I slowly became able to communicate more – albeit in a whisper.

I had a mixture of dreams interspersed with the harsh reality of being in intensive care and I assume that this happened whilst I was waking up. I don't remember much actual 'treatment' in ICU even though there must have been a lot but I do recall staff coming in to 'turn' me to prevent bedsores. I can remember times when I had to be cleaned because I had soiled myself and I was painfully aware of the cleaning process. Maybe those are the turns I remember most because I had a partial awareness. I know I often thought the turning was pretty pointless once they'd done it because I was essentially still on my back with a blanket under the other side of my body but they knew what they were doing. I wasn't able to voluntarily move much of my body – even though I didn't really seem to understand that. How did I not notice? So my body was put into different positions to keep it moving. This turning continued until well after I left ICU and was transferred to a ward.

I was still sleeping a lot and I had no concept of day and night. I know more than once I asked when the speech and language therapists were coming in the middle of the night because I needed a drink and I was told that it was in the early hours of the morning and I needed to go back to sleep. But I didn't really know where I was or what was going on so going to sleep just meant launching into a different dream. I wasn't conscious of it but sleeping didn't always bring relief as some of my coma dreams were really frightening. I felt out of control. I was totally unable to understand what was really happening.

I remember being asked daily, "Can you tell me where you are?" I don't think I actually answered correctly until the very last time I was asked. But every time I was asked before that I thought I knew exactly where I was – what a ridiculous question! But I do remember the 'fist bump moment' in my head when I finally got it right – that was on my last morning in ICU just before I was stepped down to HDU. It was only at that point that I realised I hadn't been particularly 'present' up until then.

Some dreams were believable and others were outrageously unreal. The more 'out there' ones are weirdly reassuring because, even though they felt incredibly real at the time, I know with absolute certainty now that they weren't. Chances are the more believable but unpleasant ones were also mostly not real but I can't be completely certain.

The dreams I had in delirious moments featured various people I knew but their inclusion was somewhat random and even now I'm not really sure what dictated who was involved.

My location was controlled by an interactive whiteboard which was in my room in ICU. People often came in and changed the whiteboard which sent me off to different places. I was worried about using it incorrectly even though I didn't once touch it myself – it was out of my reach. I'd never seen anything like it. It was incredibly high tech and had been provided by my friend and colleague Maria and her brother Pete. I felt very lucky that they had given this to me.

The board did exist. It wasn't interactive clearly but it was changed regularly and the first time I was really aware of it, it said *'Happy New Year!'* I remember staring at it a lot in my more lucid moments just in total disbelief that it was 2022. How could that possibly be? I didn't remember Christmas or New Year! I suddenly felt a panic that I needed Mum to bring in my knitting because I was making Christmas stockings for my cousin's young children – I'd finished one and really needed to finish the other before … I remember that penny drop moment when I realised that it was too late and she didn't get her stocking in time. I was so disappointed for her.

Before I went to sleep the South African variant of Covid *Omicron* was just starting to take hold. It was considered milder than previous strains but was more transmissible. Unfortunately, from my reaction, it seems that I caught the previous *Delta* variant as I was later told that if I'd caught Covid a couple of weeks later I probably would have caught Omicron and been fine. I was aware of this new variant and I must have gone to sleep thinking about the South African staff members at work. I can only think that it was for this reason that I went out to South Africa in my delirium because I've never been in real life. I know I felt terribly sorry for them all because in my mind they all had to go back to South Africa to their families – I'm not sure why. Then I started to worry about how on earth the nursery would cope without them because there are a few of them. I was trying to find out how long they would be gone for but I couldn't.

Next thing I knew, the whiteboard was changed and I woke up in a building in South Africa. It was hot and dusty. I could smell sand and desert. I was in my bed but I was surrounded by windows so that I didn't need to move much to see everything outside. I was hot and frequently tried to uncover myself but I just got tangled in my sheets when I tried.

> Dear Jennifer, I am looking after you today. You are on ventilator support and laying on your tummy to help the oxygen get into your lungs and to the rest of your body. We have given you sedation and paralysed you to keep you stable. You are very unwell and we are liaising with a lung specialist team. We are treating you with a high steroid dose and antibiotics. We are planning to turn you onto your back in the late afternoon. Hope you get better soon.

Outside my window was mostly desert but there were occasional trees and patches of grass with safari animals. They were just milling around on the reserve. The sky was blue and the sun was always shining – it was never dark. Maria was in charge and had arranged for me to take the trip to work out there (I didn't see her) and the first thing she did was to give me a baby black panther as a gift. I was a bit surprised but was delighted with my new baby. She said they wanted to give me something because I'd been so brave. He was tiny, my panther, but he grew very quickly and was soon a proper big black cat.

I didn't leave the room I was in so could only watch him from inside but I saw him grow into a wild animal. He joined the other animals on the reserve. I felt incredibly lucky and grateful that I'd been given the opportunity to experience the safari as it's something I have always wanted to do. Sometimes I was just in a room in my bed watching out of the windows and sometimes I was in a laboratory – in my bed still. I spent more time in the laboratory than anywhere. I was intrigued as to what they were doing in there and sat quietly watching it all.

In the South African laboratory I saw a number of people I recognized from work. They were very busy and wearing lab coats. One bustled past me and didn't actually recognise me. I was quite offended because I didn't think I had upset her but I only saw her the one time.

One of my colleagues was clearly in charge of the lab and she was extremely capable. She knew exactly what she was doing. She was experimenting on something but I couldn't see what it was and I can't remember what she had said she was looking for.

She kept on coming back to check on her experiments. One day she came in and there were no scrubs for her to wear so she went and got a hospital gown which is what I was wearing. I thought it was nice that she wasn't ashamed to look like a lesser mortal like me. She carried on with her experiments in her hospital gown. The other staff didn't bat an eyelid when they saw her wearing that.

There was also a really young guy in the lab. He was generally there on the night shift when the lady in charge wasn't there. I thought how dedicated he was – he barely moved from his station the entire time and I admired his obvious fascination over what he was doing. He was really nice and actually spoke to me – almost everyone else in the lab didn't speak to me or acknowledge me. I thought it was a bit odd but just assumed I was a bit in the way.

> Hi Jennifer, I am Sam and I have been looking after you today. You still have a breathing tube and you are on a ventilator. Your lungs are taking some time to heal so we are giving you medicine to keep you asleep. We needed to put you onto your tummy earlier (today was the 4th time) so that the gas exchange in your lungs might improve.
>
> Your mum called this evening and I updated her as she was quite worried about you.
>
> It snowed a little today for the first time this year!
>
> I hope you get well soon

The male laboratory worker was friendly and chatted to me whilst he was working – he wasn't anyone I knew. He was experimenting on something directly opposite my bed and spent all his time on it. He only left it to use the toilet (only there aren't any toilets in ICU so he had to use a bedpan!) But in between that we chatted easily. He was very short and wore glasses. He had difficulty getting on and off the stool he was sitting on whilst experimenting because he was so small. He had a pencil behind his ear and his shoes were enormous. He didn't show me what he was doing but he was using a microscope to monitor the development of something in a Petri dish.

My cousin's husband was there in the lab wearing a white coat. This was slightly different as it wasn't for single experiments. He was working on some sort of a production line with the animals and was always busy directing operations. He didn't seem to see or speak to me at all and I could only see him from a distance but he was heavily involved and very skilled at what he was doing. I couldn't work out why he had zebras and elephants on a conveyer belt. As they slid past him he treated their ailments. He had thick goggle glasses on (he doesn't wear glasses). His inclusion is quite random as I see him less than most of my family who didn't end up in any dreams that I remember.

> Hello Jen, my name is Julia and I was the nurse looking after you today. I have met you before but a while ago.
>
> Today you remained intubated and connected to a ventilator. You are on renal dialysis. We kept you sedated with intravenous medications but as your breathing was slightly faster than we would like, we added a new medication to keep you more sedated and calm your breathing.
>
> Your mother called this morning for an update and she said she will bring in recorded messages from your family and friends. I played you a few of your favourite CD's and we listened to some Christmas music.
>
> Your right eye looks a bit sore and you weren't able to close it so we closed it with some gauze and tape.
>
> Keep fighting Jen

I was then suddenly sitting on a sofa in a small stone house with a large fireplace and winding stairs where I lived by myself. It was a round house and the room itself was round too. It had a spiral staircase leading off the main room to upstairs. The room was tiny but had a roaring fire and white stone walls. It was very rustic. I lived there (briefly) but didn't go out of that room. I owned that tiny house and lived there with my dogs. It was really cosy and I loved it.

The fire was an inglenook fireplace and it was enormous – took up most of the room. I was thinking about going to bed but I don't think I moved from the sofa before I was somewhere else.

I got involved in a business with someone who I knew of many years ago from scouting. She was already doing this business and I couldn't believe that no one had done it before because it was really clever but it was also a little deceitful. It involved wearing disguises and being different people in different places. I had seen her in one place and then went somewhere else and saw her again – but she looked completely different. I had rumbled her and she then roped me into the business.

I don't remember the details but I was moving around to a variety of places as part of the role and I was in a really nice sporty car. The seats in the car were laid back bucket seats and there was a TV screen within the car that was involved in the business. I just don't remember how the car was involved! It's very frustrating to not be able to remember as this was one of my main dreams and it was quite convoluted and detailed.

> Hello Jennifer. My name is Hannah and I have been looking after you today from 8am to 8pm. I'm sorry that you have to go through this difficult time but hopefully you will be feeling better soon and recover from this illness.
>
> The main achievement today is that we stopped your sedation to assess your respiratory and neurological improvements. You were de-proned and are currently laying on your back. Your eyes are open but you are

not responding to me yet. The sedation may take a while to wear off. We have reduced the ventilator a little to see how you cope.

I wish you a speedy recovery and hope you keep on improving

Next my brother Andrew was in his garden and I found out that he had crashed Pete's (Maria's brother's) helicopter – flying it from house to house. He wasn't at all bothered and he sat with his feet up in the garden while Pete was absolutely furious … I was completely mortified and I couldn't understand why Andrew was so relaxed about it. He's normally really respectful and careful with his own belongings, let alone someone else's! This was totally out of character. He was just pretending it hadn't happened.

Pete's house was the 'Teletubby' house with the rolling meadows and the house almost completely under the ground – covered in grass. I was trying to smooth things over between them but they were so far apart that I had no chance really. He was ranting and ranting whilst Andrew had his feet up on a sun lounger, seemingly unbothered. I was watching from above but was talking to both remotely. I was stuck in the middle and I was trying really hard to smooth things over between them. Andrew is very respectful and conscientious so it's very odd that he was suddenly this character who didn't care. It couldn't be further from his perfectionist personality!

Hi Jennifer, I am Carla your physio. You are asleep and still have the breathing tube. I have checked your breathing today and moved your arms and legs.

Keep fighting!

The whiteboard unexpectedly then took me to the bottom of the ocean in a submarine. Other people were there and they were all in the military. They were wearing camouflage gear. They were scavenging across the bed of the ocean for treasure and bringing back items they had found.

It was the first time I had been and I was amazed that they were bringing back electrical goods and high value treasure that they had found. The first thing they brought in was a washing machine! They brought it all in to show me but made it clear that they didn't know why I was there. I was quite obviously in the way here too but they were all pleasant to me. There wasn't anyone familiar there.

The inside of the submarine was dark grey metal all over. There were windows but all I could see through them was dark water – maybe the odd fish. I wasn't sure why everyone else was going out and I wasn't but I was really tired and felt like I was better off staying in bed anyway.

My bed was in a massive room and people were in and out all the time. Someone was standing at a computer screen directing everyone and the team were coming and going with their treasure. It was quite high tech and there were screens and computers all over the walls with people walking between them, tapping them and inputting information.

No one was wearing any scuba gear and none of them needed masks or oxygen so I thought they must all be really good at holding their breath! But they were military – they could do anything. They also didn't appear to be wet when they returned. I couldn't quite work it out.

> Hi Jennifer, my name is Ava and I have looked after you overnight. You are still on the breathing machine so you are asleep. Your kidneys needed some help so

you have a filter going to keep them clean. You had a stable night tonight. I hope you have a good day.

I also spent a lot of time in the local chemist with the whiteboard – only I was upstairs in a private flat where they were all sitting on the floor unpacking and preparing for a business opening. I'm not sure what the business was but they were all very busy and focused. They were terribly excited about it and I really wanted to be involved. They were chatting easily with each other as they worked. Again, they didn't seem too sure why I was there and I of course had no idea but it was a bit like they were tolerating me because they had to. There was no furniture, just cushions on the floor and they had stock all over the place that they were trying to sort out. My bed was on the side so I could see the whole room and everyone in it.

At one point they said they were worried about the atmospheric pressure in the room as it was a new room. I told them that it was okay because I have a shunt in my head for long standing intracranial hypertension (I have two) and I can monitor the atmospheric pressure for them. I then started telling them I could actually control the pressure with my shunt so they really didn't need to worry. I felt important and assured. But they looked at me like I was mad until I suddenly realised what I was saying and quickly stopped talking. I explained that I could maybe sense it with my shunt but not control it – in this version of reality anyway! Even though atmospheric pressure does undoubtably affect my shunts, I really can't do either. I wonder if I did actually say this to someone …

Hello Jen, it's Julia again. I've been looking after you today. You are still sedated and connected to the ventilator via a tube in your mouth. You are also still on dialysis.

Your eyes are still sore and we are putting antibiotic drops and moisturizing gel every 2 hours. The doctor asked us to keep your eyes covered for most of the day.

Apart from this, we changed the settings on the ventilator this morning but you did not seem to agree with this so we have tried to revert back.

I played you the CD's your mum brought in with some lovely messages from your family and friends. You are obviously loved.

I hope you have a settled night

Whilst this was happening upstairs in the chemist I kept getting really tangled in my blankets. I stayed uncomfortable for quite a while and when I tried to call out to someone I couldn't get their attention because my voice wasn't working properly – most likely because of the tracheostomy (in fact eighteen months later it still has very little volume and vocal range). I was trying everything to get them to notice me and I was embarrassed that I couldn't get any sound out. I don't remember having a call button at all but of course I would have had an assigned nurse with me most of the time.

When I finally did get their attention, I really felt like a bit of a nuisance when they were so busy. I felt like they must all be rolling their eyes at how annoying and pathetic I was. I remember the relief when they came over and untangled me though because I finally felt comfortable again – but the next thing I knew I was tangled again. It got to the point where I didn't ask anymore because I knew I would be in a muddle again almost immediately. It was so uncomfortable being caught in the sheets and there was nothing I could do about it by myself. I don't know how I was getting tangled in the first place as I didn't have much movement but I must have had just enough to get myself in a knot with the bedding. I was probably a little uncomfortable generally.

The lady in charge at the chemist had dark curly hair and reminded me of Mum with her perm when I was little. She was nice to me and did help me to untangle myself once or twice. For some reason I felt embarrassed to ask her to help me as I thought she would be angry but she wasn't. They were all around me working hard and I was laying in bed doing nothing but asking them to come and help me, disrupting what they were doing. I felt so guilty for bothering them.

This seems to be a recurring theme – feeling in the way. It seems I felt like that the whole time. In the way and a burden on everyone else. I felt guilty for needing their help when they were so busy. I felt like a small child. Totally helpless and dependent on other people. A small child surrounded by busy adults who had no time for me. I couldn't make my voice heard so even when I did need help I couldn't get anyone's attention. I would have had an assigned nurse but I didn't know that.

While I was in the chemist, there was a phone call and a nurse took the call next to me.

"Jennifer, it is the police," she said. "They have received a call saying that Jennifer Higgs is being held against her will and they want to know if you are ok?"

I was shocked but suddenly realised it was being suggested that it was me who actually made the call to the police. I was upset that they thought I had caused problems when I wasn't doing anything wrong and when the lady in charge came over to talk to me about it, I gave her my phone so that she could check the last number dialled to prove that I hadn't called them.

I had my phone with me but, as I had absolutely no idea how

to use it, I couldn't possibly have called the police. It was like my brain and fingers were still disconnected. That and being in ICU meant I had someone with me all the time. But I was so upset that they thought I had betrayed them. I was also panicking that the police were going to turn up and I felt insanely guilty even though I hadn't actually done anything. The lady in charge confirmed that the call hadn't been made from my mobile phone and I was relieved that they knew it wasn't me but couldn't understand who on earth would have called the police. Although it sounds farfetched now, I truly believed for quite a long time – months in fact – that the police had actually been called!

Funnily enough on revisiting ICU I found out that patients do actually often call the police because the medical staff are generally doing horrible things to them and they feel like they have been kidnapped and that they are being tortured. I didn't as I couldn't use my phone but it's interesting that this was one of the dreams that bothered me most when I left ICU because I felt like it could have happened.

Hi Jennifer, it's Carla here from physio. I was with you today.

You now have a tube called a tracheotomy! This is a good thing because you are managing some time off the ventilator.

Your chest was cleared this morning with me to prevent any nasty infections! You have changed ventilation medicine today to help your lungs to expand and prevent reduced air entry. You are tolerating this really well! You are now starting to cough during treatment!

Keep fighting.

At some point the upstairs of the chemist then became A&E – maybe that was what they were preparing for before. They had decided to move it to there from the hospital and they were getting ready for the first patients to arrive. They were waiting at the door and I was watching from my bed opposite. Sliding glass doors had been installed and a number of trolleys had been set up to accommodate patients. The staff standing by the door were encouraging people to come in as if it was a shop.

Suddenly there was a man sitting opposite me and he had come in with a patient who had been taken for a scan. He was alone and I can remember trying to work out why he was allowed in when I was by myself without my family. He was very anxious and seemed worried about his relative but was watching me and seemed to be trying to pretend that he wasn't – looking over and then quickly looking away when I caught his eye.

He was eventually joined by the patient. I didn't see who they were and don't remember much about them except that they had a lot of nasty wounds on their face. I don't know if the lady opposite had a visitor one day as she was terribly ill or whether I dreamt him entirely. Or whether I actually was moved and was temporarily opposite someone else. The person I saw certainly wasn't as sick as the real lady opposite so it was very strange.

Hi Jennifer, it's Carla here again.

You are doing really well on the ventilator and trache. You are managing to maintain better oxygen in your body with being able to reduce the settings a little.

Your chest was cleared by me again today and I spent some time moving your muscles to prevent stiffness.

Keep going!

On one day I noticed the doctors crowding around the lady opposite. They were talking animatedly about taking her for a scan and trying to decide which of them could take her. Wherever she was going, she was going in a helicopter and they all wanted to go with her on the exciting trip.

They eventually decided and two of them came back wearing red boiler suits like the HEMS team. They wheeled her away and off they went. Suddenly, I was on my own. There was no one around and after a short while, I started to get a bit anxious.

I gradually became aware that there was an event going on in the grounds of the hospital. It sounded like a fairground or a fete of some description and I could hear everyone having a lovely time just outside my window. I figured that the nurses who were meant to be caring for me must have gone to this event. I was glad they were having fun but I really wanted them to come back!

Realistically, I know that I wouldn't have been on my own. I was in ICU. The care is pretty much one to one. The lady opposite may have gone somewhere but I'm not sure where. Maybe she just went for an ordinary scan and it just seemed like a long time. She did come back but I just remember her being there. I don't remember her actually coming back!

> Hi Jen, my name is Alma. I'm the staff nurse looking after you. You have been quite stable overnight and you appear comfortable. You managed to slightly wiggle your finger at the beginning of the night.
>
> You slept for a few hours and I have had to tape your eyes closed to prevent them from becoming dry because they are sore. You had a settled night.

Sometimes the whiteboard took me off onto a boat – usually our old boat from when I was a child, and I was squashed in the front in a bed that was uncomfortable with a curtain in front of me. It wasn't terribly big and I had to curl my body to match the curve of the bed. There were various other cabins on the boat and they were occupied by other people, surrounded by curtains.

People kept coming in to 'turn' me and clean me up and it felt like they were doing it at all hours, whenever they felt like it – I'm sure they weren't and I'm certain there was always a good reason. But to me it sometimes felt like they wanted to punish me and chose to turn me then went off laughing about me and how useless I was.

At one point two staff were left in charge of me and I thought they were having a relationship and were turning and cleaning me out of spite because I had rumbled them. One was ginger like Gary from Coronation Street (he was there a lot) and the other was a pretty young girl. They were both nurses. They had a good laugh about it afterwards but she helped to roll me over and he was cleaning me. He sprayed some sort of cold foam and then they put me back onto my back – with a blanket under me so I was slightly lifted on one side.

> Hi Jen, it's Kayla here from physio. Today you sat up on the edge of the bed. You did amazingly!
>
> We looked at some pictures of your dogs (they look very cute) and of you with Lynsey. We have stuck them up for you to see. Nadine also put some music on for you to listen to after all your hard work.
>
> Keep going! You've got this!

The physios on ICU continued to see me every day. They got me up – I wasn't too impressed by all accounts. On their first visit they told me that they were going to sit me up. Well that's easy, I can sit up! To my horror it took all three of them to get me to a sitting position with one person kneeling on the bed behind me to keep me upright and another sitting on the floor holding my legs that she had just lifted around to make my body sit. They had my phone and one of them asked me to smile while she took my picture to show my mum. I tried my hardest and they all cheered and told me I was doing brilliantly. I needed to lay down again after that. I suddenly felt exhausted and couldn't stop myself from falling asleep. I have this photo and, even though I can see it, I just don't believe it. Even years later it feels like a really scary dream or a story I'm hearing from someone else rather than something that actually happened to me.

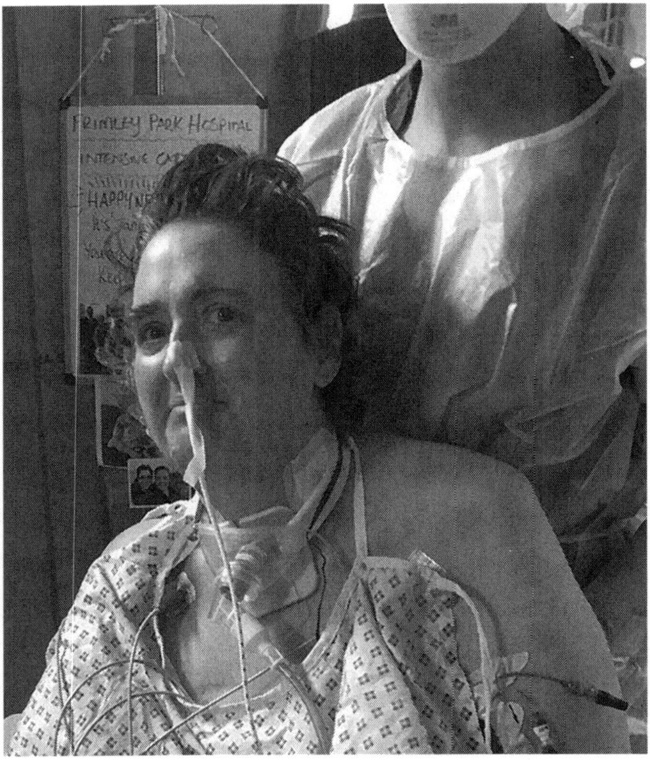

The whiteboard and the tracheostomy

8

REACHING OUT

One of the physio's Nicola was really kind and spent a lot of time with me. She promised that she would come and help me call Mum. I hadn't forgotten my family but it wasn't until we started talking about calling them that I realized they hadn't been there! Despite the code for my phone being written on a sticky label on the back, I couldn't open it. It was incredibly frustrating because it just kept locking. My fingers felt like they belonged to someone else! So Nicola brought me the ward phone and dialled Mum. I spoke to her for the first time.

I don't remember much about the call. I know I told her I was desperate for water and they wouldn't give me any! I know she also said my brother had given her Lego for Christmas and I smiled because I had done lots of Lego but she never had. She said it was Lego flowers! I couldn't really picture that.

The physios came back again. It really didn't feel that long since they were last there. They told me that they had a chair for me to sit in. They showed me a posh looking recliner chair and I thought how lucky I was that I got to sit in such a comfortable chair.

> Morning Jen, today you sat on the edge of the bed with the help of me, Carla and Kayla. We then hoisted you to the chair.

I took a photo and sent it to your mum (after a slight technical problem).

You are doing so well! I left you listening to Aled Jones!

They put me into a sling and hoisted me up. As I moved away from the bed there was a splatter and I was horrified when I looked down and there was mess all over the floor. I was not aware that I needed the toilet so it couldn't have come from me. Could it? They hoisted me straight back to the bed. I asked them what that was and they told me not to worry. They then proceeded to clean me (and the floor) up. The physios left and the nurses finished cleaning me then before long I was back in bed. I drifted off to sleep again.

The next time I was hoisted it was by an agency nurse called Tim and a care assistant. Tim was looking after me on that day and he seemed really nice. He introduced himself and was really rather good looking as well as very kind. They put me into the sling and hoisted me up. I wasn't very comfortable and felt a bit unsafe so I tried to let them know but I couldn't speak loud enough. They just kept on hoisting me up until I was so high I could see the top of the curtains! I started to fall out but couldn't scream or make enough noise to communicate it.

Eventually they realized and brought me back down again. I was so relieved to be back on the bed!

"That shouldn't have happened and I'm so sorry," said Tim. "Do you mind if I come back and take a statement from you later?'"

I thought that was a bit odd as it was partly his fault but I said that was okay and he went away. In actual fact I think I probably would have agreed to just about anything as he really was very handsome and I had developed a bit of a crush! He didn't take a statement or mention it again.

Given that the staff in ICU are highly skilled, I doubt that this actually happened but weeks later I still very much believed that it did to the point where I was telling the staff in rehab all about it before I suddenly realized that it might not have happened.

> Morning Jen, it's Carla here from physio.
>
> Today we got you to the chair using the hoist which lifts you up in the air. I ensured you were as comfy as you could be as I know you're not the biggest fan of it!
>
> You are doing so well! Keep on fighting

I was hoisted a lot after that. I didn't ever like being put into the sling and then lifted by the hoist. It felt scary and uncomfortable. I'm sure I moaned and groaned about it a lot. I didn't feel safe at all and it was certainly completely out of my control.

When you are first lifted up in a hoist, you sort of have to fall backwards into the sling. Once you know about this fall it's fine – even fairly comfortable for a short while, but when you don't know or you are a bit confused, it really feels like you are falling backwards. Your whole body tries to tense ready to hit the floor and then you're just waiting for it to happen. I think this may have been the problem when I felt like I was falling with Tim. I like to think so anyway.

> Morning Jen. I got a lovely wave from you today plus a nice smile.
>
> Today myself, Carla and your nurse helped you to sit on the edge of the bed again. You did really well and used your left arm a little to support yourself.

> We will come back and do some physio again later. I will also help you to call your brother. You are very tired after all that physio!

Nicola told me she would come and help me to call my brother. I remember a little more about this call than the first one with Mum. In my head, he was on a recliner chair in the garden working on his laptop when he answered the call. He was surprised to hear from me I think. I can't remember what we talked about but it seemed that he was very relaxed on his sun lounger – in January! I think he was a bit taken aback by the call.

It was so lovely to hear the voices of my family but it also emphasized how alone I really was – even though they were all there rooting for me from a distance. I wasn't able to just text like I would normally and they couldn't text or call me.

The physios on ICU were incredibly positive and encouraging. Even when I was grumpy and uncooperative. There were three of them and they all seemed much younger than me. In my mind I was trying to work out if they were my cousins. I couldn't understand why they were there. I thought one of my cousins could easily be a physio – I wasn't sure, but I didn't think all three were. It seemed they were right there though so I went with it until I was conscious enough to actually ask them if they were related to me. They weren't obviously but they said it wasn't unusual for patients to mistake staff for family members where they were desperately gripping onto something familiar and trying to make sense of the environment and people. At this time I was still unable to see anyone I loved as visiting wasn't allowed. What I desperately needed was familiarity and comfort from family. It was very difficult not having that and my brain obviously tried incredibly hard to bring that familiarity back in whatever form it could.

Hi Jennifer, I'm Nicola and I'm part of the physio team. Part of my job is to move your arms and legs whilst you are sleeping.

I called your mum today to find out some information about you. She told me about your two dogs. I know you love music so I have put some of your favorites on for you.

Keep fighting Jen. You are doing such a great job!

Staff caring for me were wearing masks all the time so I was only able to see their eyes. It's amazing how much masks distort your view of a person. It's funny, when someone familiar wears a mask you know what they look like underneath and your brain sort of fills in the blanks. We all got so used to seeing people in masks during Covid lockdowns and in hospitals and I never realised that, when you don't know them, you actually subconsciously form a version of their face in your mind. I thought I knew what the people caring for me all looked like from just seeing their eyes. It wasn't until much later in rehab that I first saw the faces of those people who over three months I'd spent all of my time with – some of whom had been involved in my most personal care. Until I heard their voices I genuinely had no idea who they were when their mouths were uncovered! I wouldn't recognise any of them in the street. The lower half of their faces was completely unfamiliar and made them look like totally different people from the people I had pictured in my mind.

Everything was so jumbled in my head that I couldn't think about normal life – my mind was taken up with trying to get water, sleeping, dreaming and trying to make sense of everything I was feeling and seeing. The one strange thing, which is also a blessing, is that I don't actually remember any pain. Just complete confusion. I know I was on a number of very strong medications.

Unfortunately I do need to write about toilets. A strong feature of my recovery until at least April 2022. It's uncomfortable to talk about or even read about but I can confirm that it is even more uncomfortable being stuck in the nightmare of either having to use a bedpan or soil yourself and then being unable to clean your own body – so be prepared to squirm or skip these paragraphs.

I asked repeatedly on ICU to actually go to the toilet and I was told that I was in intensive care – of course there weren't any toilets! I didn't believe them and asked if the staff also then used bedpans. I was told they did – of course no one actually would have told me that! I felt like I was stuck in one room and that there was nothing outside of that room. Like a sealed pod.

I kept being asked if I wanted any music on. As far as I know there was no TV so I must have been either sleeping or trapped in my own mind because I didn't feel especially bored. I was worried about disturbing the other patients with music so I didn't often say yes but in the times I did, I think it was the same CD being played. It was music I liked so I knew it was from home but it was quite repetitive after a while. It didn't seem important though. I was happy to take it or leave it. It was strange not needing to be constantly entertained. My thoughts were distracting enough.

I spent a long time observing the patient in the bed opposite me. She was all I could see during awake times. She had a nurse assigned to her who looked strangely like someone I worked with and she was very poorly. I felt a bit like I was watching a film. On the odd occasion that I tried to get the attention of the nurse caring for her, I wasn't able to which made it more like something I was watching rather than participating in. How strange that I didn't realise that I was in exactly the same position as her! She features in almost all of my ICU memories.

Twice I remember having my hair washed by a lovely healthcare assistant (I thought she was a doctor). The first time I wasn't really aware and, although I was in my bed in ICU, I was somewhere else in my head. The second time I remember Margaret took a lot of time over it, drying it afterwards and I kept apologizing for taking up her time. I couldn't believe a doctor was taking the time to wash and dry my hair and kept thinking she must have more important things to be doing!

> Morning Jen! Well what can I say?! I've come in at 9:30am and you have already had a hair wash and blow dry! You look like a different person! Margaret is giving you a proper spa treatment today, massaging your hands and moisturizing your skin.

When the physios came in they were shocked that my hair had been washed and dried – I think it might have been dried quite fluffy. I can't really remember how she did the hair washing but I do have a vague memory of the bed tipping backwards because I was scared I might slide off.

Margaret must have had my head in a bowl but I felt like I was hanging off the bed and the water was showering all over the floor. I missed those hair washes once I was out of ICU until I was able to do it myself which was a considerable while afterwards. It was very difficult even then because I couldn't shower alone (I tried!) and my arms were weak – I couldn't move my right arm for quite a while.

Suddenly, I was being moved to a different part of ICU. But it was in the middle of the night. The patient in the bed opposite was going too and she went before me. Margaret was very angry and reassured me that I wouldn't be going if she had anything to do with it. She was furious that they were moving me during the night. She was packing my things and I couldn't understand why she was doing that if I wasn't going but she was really banging things around in temper. She hid things and moved them around so that it was impossible for me to go, and then disappeared and hid in her office (I still thought she was a doctor).

I didn't go that night but I was really on edge waiting for them to take me and, as Margaret was now hiding in her office she couldn't reassure me. I was desperately looking for her because I was certain that she would look after me but she was gone.

A lady from my choir suddenly arrived onto the ward and she was the matron. She was called because they couldn't find what they needed to move me and she stood with her arms folded across her chest – very angry that someone had hidden things on purpose to stop me from moving. She was wearing an old fashioned matron's uniform with the white cuffed short sleeves and the starched frilly hat. I think parts of this dream could have happened as I did eventually move to another area of ICU and again to HDU – although it was unlikely that this was happening in the middle of the night. My concept of time was very unreliable.

I was terrified by the prospect of moving and didn't want to leave where I was. I had been in there for a long time and it felt safe. Everyone who looked after me was kind and I was really frightened about where I would be going to next. Fear of the unexpected. I was also still very confused and disorientated so it was quite difficult to comprehend changes. Who would be looking after me? Would I be able to see my family? Would I be on my own?

Morning Jen. I was so pleased to come in today and see you awake.

Zoe and Jo, the physios, gave you a voice box on your tracheostomy for 15 minutes. You said 'hello'! Your first word!

You are doing so well. Later we are going to use a 'tilt table. We will slide you over onto it and whilst you are strapped onto it, we will slowly stand you up.

I look forward to seeing you standing!

Then the cleaner walked across the ward and it was my uncle. Every time he walked past me I watched him intently trying to work out if it actually was him. I knew he was in Wales and was unlikely to have been working (given that he is retired) but I convinced myself he could just be doing a little job on the side. He had slightly darker skin than my uncle which made me doubt myself but I couldn't look at him without thinking it was him. I knew I had asked the physios if they were my relatives and I remembered that they weren't. It made me question it with myself and I wasn't sure if I had made the same mistake. I think I was aware that I was a bit confused by this point. I watched him going backwards and forwards and really wanted to ask him if he was my uncle but I was too scared.

9

PARCHED WITHIN WATER

As the delirium started to ease up and once I was a little more awake, I was being weaned off the ventilator. I had no idea I was actually on a ventilator because I had been given a tracheostomy. It was a total surprise to me that I even had it because I hadn't noticed! You'd expect to realise or have some sort of idea that there was a hole in your neck and that a machine was attached for much of the day breathing for you but I genuinely had no idea. The staff kept telling me that I had been off the ventilator for short times, gradually increasing.

Eventually, once I was managing without the ventilator, the tracheostomy was removed. I can remember a lot of talk about it maybe happening and it felt like they were talking about it forever but not actually doing it!

Once it was decided to take it out I remember them gathering around my bed to remove it. It felt like a huge fuss and I couldn't understand why. Of course removing a breathing tube that connects the outside world to your windpipe is a big deal but I don't think I really understood that I had been unable to breathe for myself at all. It didn't feel real. I was also unable to sit up by myself and so I needed support behind me whilst they did it.

> What a start to the day! Jen, Kayla removed your trache!

So, whilst you were in bed myself, Kayla and Carla prepared you for it to be removed. It came out really easily when you coughed. You wanted to see it too!

We decided to give you a break and let you rest for a while then we are going to try and FaceTime your brother!

You are amazing Jen

When they took it out I remember people around me – reassuring me, holding my hands and holding me upright. I don't know what I expected but the little tube that they removed wasn't it! I don't think I reacted, flinched or anything. Didn't even feel it! They showed it to me and it was just a plastic tube about 10cm long with a bend in it. That tiny plastic tube actually kept me alive. Even now it still feels totally unreal. I don't remember them stitching the hole in my neck and once I was a bit more conscious, I took a photo of it. It was like a vertical slit in my throat, around an inch long. Now it is just another scar. Adding to an already impressive collection!

Once that was out they started to talk about eating and drinking again. I didn't realise that I hadn't been eating or drinking for this whole time but I did have a feeding tube up my nose and into my stomach – and I was thirsty. Oh my goodness I was thirsty!

The speech and language therapist said I could start trying some fluid but because I had a hole in my throat there was a danger of me inhaling anything I tried to swallow and this could cause a chest infection. She said I could start having water and I was really pleased until she went to get the thickener and I tried it! The thickener is a powder added to liquid to give it the consistency of wallpaper paste – and to prevent water finding its way into the lungs.

It was utterly vile.

Things got harder after that because I was so thirsty. I felt permanently dehydrated. I didn't have much voice and communicating was difficult so I know I got very frustrated at times trying to ask for water. I couldn't believe they were keeping water from me when I was so desperate. Of course there was good reason. To me it just felt like everyone was being really cruel and withholding what I needed most when I was desperate for it. I didn't feel like I was asking for much. Just water.

Whenever I asked for water they told me that I needed to wait for the speech and language therapist to come back and see me so that they could evaluate my swallowing and decide what I was allowed. I think I probably asked every single person every time I saw them when speech and language were coming back. I started to wonder if they even existed anymore because they always seemed to be coming 'tomorrow'. Sometimes I would ask and they would tell me that they'd been already. That was the most confusing thing because I had no recollection of it – and I still wasn't allowed water so in my head I'd missed my chance. It was like a one time deal and somehow I kept missing it. This just made it more and more frustrating.

A doctor came in on the ward and imagine my surprise when he was gorgeous Ethan from the TV programme *Casualty*. He felt sorry for me because I was desperate for water and I begged.

"Oh go on then," he said and then he gave me a whole bottle of water! I was thrilled! He told me to just take small sips and I did as I was told.

When speech and language finally did come, they said I was allowed three 100ml portions of thickened fluid per day and wasn't able to have any more than that. They took away the bottle of water I'd been given – "No more sips," she said. I was devastated as, once again, I was left gasping for water.

Considering my enormous thirst, the three small portions just didn't feel like enough. I was allowed apple juice or coffee as long as it was thickened. I can honestly say that thickened fluid is disgusting! No matter what drink it is mixed with! Even now, I can no longer stand apple juice because that was the best of a bunch of bad options – and I had it a lot! Quite often fed on a teaspoon.

When the physio team came in I asked them where my bottle of water was. I had completely forgotten speech and language taking it away. They were confused so I told them the doctor had given it to me. This caused a huge commotion! "It's not his job to do that!" "Doctors have no idea!" "Even if we find it you won't be getting it back!" they said.

I was devastated.

The physio team went off to speak to the doctors who said that not giving water was ultimately inhumane (exactly what I was thinking – imagine that …!) and, as a result, they rewrote their policies surrounding water for ICU patients who were waking up. They designed water packs which had pipettes of water that patient could drink whilst they were waiting to be assessed by speech and language. I couldn't have one because I had already been assessed.

Given that there was good reason why I couldn't have liquid and ICU policies must be very longstanding and effective, I suspect this didn't happen and I don't think any doctor would go against the specialists. So, lovely as he was and based on the fact that he isn't actually a real doctor, I think I dreamt Ethan!

10

THE EYES BEHIND THE MASKS

My hospital bed was moved down to the basement and the whiteboard transported me there. I was being looked after by someone who I used to work with. Philippa was really strict and kept telling me off. I was still only allowed 100mls of fluid mixed with thickener three times a day and she refused to give me anything other than that – some nurses weren't so precise and sometimes I think I got a bit more because I was constantly thirsty. She did find that I had a bottle of iced latte in my bag which must have been from when I was on the ward and she thought that might be okay with thickener so she made it up and gave it to me to try. It was actually okay to start with and I remember her saying that she wouldn't mind having some too. I was using a cup with a lid on – like a toddler cup, and she had to help me to drink it. I wasn't allowed the whole bottle – strictly 100mls. The thickener kept on blocking up the spout. It was very frustrating because I was so desperate for fluid! I never saw the rest of the coffee so didn't get to have it again but I wasn't particularly open to the thickener by that point anyway.

> Morning Jen! The neuro team are coming again today to do some rehab with you. You are doing really well.
>
> Best wishes

The basement was cold and eerie. Everything was dark and there were big metal posts holding up the ceiling. A bit like an underground car park. It smelt damp and I couldn't see any daylight at all from anywhere. Mine was the only bed so it was just me and Philippa there. It was isolating and scary. I felt very vulnerable and shivered with the cold.

We moved back to my cubicle next and Philippa was still my nurse. When she turned up I was gutted because I just didn't feel like another telling off! I was starting to get irate with her because she was so strict and unpredictable. In our working lives, Philippa came in after me and tried to take over the role that I was doing even though she didn't have the skills and knowledge. She was intimidating back then and one of those people who always seemed to do things perfectly on the surface. But I was the one running round clearing up her mistakes even if no one really saw that. So for her to be authoritative with me as my nurse didn't surprise me.

There was a crowd of doctors outside my curtain and they were being really rowdy. I didn't like it because it was too noisy and I was trying to get to sleep – it was the early hours of the morning after all! It turned out they were actually having a party! Outside my cubicle!

Philippa got really cross and went outside telling them that they needed to take their party elsewhere. "Would you normally stand outside someone's cubicle and make so much noise?" she said. They replied that it was fine because they'd finished anyway. They were so arrogant! They left and I went to sleep. I then felt like she had protected me and had my back but I was also a bit wary of her still.

Philippa was quite confusing because she would be really stern and mean sometimes then would soften a bit and be nice before going back to being stern. I found it difficult to anticipate how she was going to react to me. I really wanted her to like me and enjoyed her softened moments but I was really uncomfortable being looked after by her because she made me feel stupid. Especially when I was asking about toilets in ICU. But I had no idea what I was doing or where I was. She didn't seem to understand that.

Philippa kept on trying to get me to use a bedpan and I was refusing. She said the only option was to either go in my bed and be changed or use a bedpan. In hindsight I can't even imagine how I would have used a bedpan as I couldn't sit up unaided. It's like trying to potty train a giant newborn baby! At one point though I really had to go so I tried to use the bedpan. I was stood up in a room behind a curtain and I quickly used it then everyone applauded because they knew I was desperate. Obviously this wouldn't have happened quite like that as I couldn't sit up and definitely couldn't stand! But Phillipa covered and took away the bedpan I had used and then I had to be cleaned. Whatever actually happened at this point I'll never know but I'm probably best off remembering my version of events as it's likely to be slightly less embarrassing!

> Morning Jen, lovely to see you today. You are awake and seem raring to go!
>
> Today we sat you on the edge of the bed again and did some exercises. Then we hoisted you to the chair. The Doctor was watching your progress and was very happy!
>
> I see you have a little scrapbook that I think your mum made. We will look through it together in a minute.

As I was watching life in ICU I could see bags and bags of

fluid being brought in every day. It was like I was surrounded by water that I couldn't drink. Had I been a little more awake I would probably have realised I was dehydrated and asked for more fluids in my drip but I wasn't and didn't. I was desperate to get one of those bags just to get a drink of water. It was like torture seeing so much water and being unable to have any.

The staff brought me sponges soaked with water to wet my mouth and, if they gave me any opportunity, I desperately sucked out the fluid. They told me off for that a lot! I know I had a row (a very quiet one) with one of the senior nurses and I asked her how she was intending to stop me from walking over to the tap (which I thought I could see) and getting some water for myself. She was very kind and said she hoped I would listen to her and not try that. I had no idea that I wasn't able to stand or walk and I was getting really distressed.

One of the nurses came over with some small bottles of water. She said she couldn't stand by and see me denied a basic human right so hid three bottles in my bedsheets. She told me to get them out when no one is looking and have a drink.

Things got busier and I didn't get a chance to find the bottles but when I finally did, I couldn't find them. I looked everywhere around my bed and my sheets but there was no water. As I only had limited movement this would have been extremely difficult! I was upset that I couldn't find the bottles and I dreaded that nurse coming back in case she was angry with me.

She was furious.

She came back dressed fully in black so that all I could see was her eyes. Along with a team of others who were also dressed from head to toe in black and they were helping the lady opposite me to escape. She just stood staring at me through the sides of her eyes which were like slits. She didn't speak, just stood there. I started to feel really uncomfortable because she looked dangerous. Gradually the focus seemed to shift from the patient opposite and I realised, to my absolute horror, that they were there to kill me.

They were all floating by now and were just moving around me and staring. They didn't touch me. I became so frightened that I called Andrew in a complete panic and asked him to come and get me because these people were dangerous and I was terrified. He never came.

Can you imagine being totally alone, vulnerable and weak, and the only people around you are threatening to end your life? Not talking to you, just circling around dressed in full black with only their eyes showing. And when you reach out for help from family they don't come? It's truly terrifying. I can still feel that terror in the pit of my stomach.

No calls were ever made – and I was unable to use my phone so this must have been a delirious dream. But it was a dream that still haunts me now.

On revisiting ICU I discovered that this type of dream isn't unusual. Staff were in full PPE – including helmets and gowns. I would have only been able to see their eyes. They were, of course, doing things to me that were unpleasant so I may have interpreted this as them wanting to end my life rather than the total opposite. This particular incident has been revisited frequently in dreams and flashbacks I have had since leaving ICU and beyond. It is certainly one I will never forget and two years later I remember it like it was yesterday.

Morning Jen! It's Carla here from physio.

Yesterday the neuro physio saw you and reported that you are improving in strength.

Today we got you to the chair using the hoist and you

> spent a short amount of time sitting up. We made sure you were as comfy as you could be. You did get quite tired so we hoisted you back to bed to rest.

Tim, the agency nurse who had hoisted me previously, came back and was caring for me. He was good to talk to, joked with me and seemed to be on my wavelength. I really liked him. He was attentive and really made me feel like I was important. I told him how thirsty I was and he asked me what I liked to drink. I told him I liked coffee and hadn't had one since before I became unwell and he said he'd see what he could do.

The next day he was back but he wasn't assigned to me. I was so disappointed. But he did pop over and he told me that he would get me a coffee when he went for his break. I waited a while and, sure enough he came along with coffee! I imagined he'd gone into town and got me a costa but he most likely made it in the staff room!

The coffee was in a sippy cup like very young children often use. I was pretty horrified when I saw it! But I wanted the coffee. Unfortunately he put thickener in it which took it from my favourite drink to my nightmare. He was so excited that he'd got it for me though that I drank it and didn't let on that it tasted horrible.

I then overheard him talking about me saying that he was worried about my mental state because I kept on asking when speech and language were coming – even when they'd just been! I then felt like he'd betrayed me, even though I was probably doing exactly as he said. I was really upset with him and refused to talk to him, choosing instead to turn my back and pull the covers over.

I don't know if 'Tim' existed or not. After this he became a doctor and then an IT expert. He never looked after me again. I just saw him doing various roles around me but he didn't talk to me at all. I started to think he was a bit arrogant as he seemed to know everything about everything!

11

FROM NEGATIVES TO POSITIVES

Once the tracheostomy was removed and I had finally started testing negative they began to talk about a special visit with Mum in a garden they usually use for patients who are terminally ill to spend time with relatives. They said it was highly unusual but that I had been in ICU for so long that they thought it would be good for me to see Mum. I needed three negative tests though and so did she. They were very strict on this – and to me it felt like it took a month but finally the negative tests were done and I was able to see mum. It had been a long time and I was relieved to be allowed to see her – and very grateful. I was also a little anxious and I'm not sure why.

> Morning Jen, I've spoken to the consultant today about getting you to the garden to see your mum.
>
> Hopefully once you have had three negative Covid tests you can see her - everything crossed for you.

Mum was told before she came in that I wouldn't be me anymore and to expect a very different person but in my mind I was still me in every way. Actually Mum was relieved to see the real me too even though I was very quiet because my voice had been affected by the ventilator and the tracheostomy.

I was taken down by two nurses on my bed along with a lot of equipment. I don't remember them taking me there or wheeling through the hospital but I do remember crying when I saw Mum in the garden room for the first time. It had been weeks and I had been so alone through all of it. For her it had been worse because she was aware of every moment. I was asleep for most. She was pleased to see me but probably quite overwhelmed – particularly by my caterpillar sized eyebrows which hadn't seen a pair of tweezers for a very long time! All that time she had thought she wouldn't see me again and there I was. I can't imagine the various emotions she would have felt – and of course hidden to protect me.

We chatted (quietly) for while before the nurses needed to take me back. Thinking back, considering how many sick patients they had, for two nurses to take an hour out just for me was incredibly kind. It made a difference being able to see the person I am closest to in the world after being so afraid and alone. I remember them watching and being a bit emotional themselves. I still had no idea how sick I had been.

I was very weak after so long laying in bed without moving and I wasn't able to do anything much for myself. I was fed through a tube up my nose and into my stomach and was totally unable to feel or move my lower left leg and my right arm.

The physios came regularly and wanted to get me sitting up. They put me into the hoist and lowered me onto a very plush looking chair. I thought it was a bit strange that I didn't seem to be able to do it for myself. It felt like my arms and legs were a bit disconnected. They told me I could sit in the chair for about half an hour and then they would come back and get me back to bed. Half an hour seemed like quite a short amount of time to be out of bed and I couldn't work out why it wasn't for longer.

But very quickly I was in so much pain! Everything hurt! I felt like I had been sitting in this chair for about four hours! It wasn't as comfortable as it looked! I asked when I could go back to bed but the nurse told me I must wait for the physios.

Eventually they came back – and they said they had only been half an hour but I thought they must have lost track of time because I had definitely been in that chair for longer than that! They hoisted me out and back into bed. I was very grateful. The daily chair sitting continued onto the ward and the length of time I sat there got longer – and didn't feel like it was getting any easier.

I had to be strip washed everyday and it involved turning me in the bed to get to different parts of my body. I was naked and very aware of that as I became more awake, but they did generally try to keep me covered. As I was really weak I wasn't able to turn my own body or lift my limbs so had to be turned by staff as they washed me. I dreaded them coming to do it because it seemed very intimate and I didn't really understand what was going on. Once I got a little stronger I was able to help with the turning by pulling myself over using the cot sides of the bed but at this point, still in ICU, I had no strength and had to just let them do whatever they needed to.

The team of neuro physios continued to work with me as I became a little more awake. They were stroke specialists and they were focused on getting me more mobile ready to step down to HDU and beyond. They also worked on HDU so would be supporting me there too.

Breathlessness was a significant problem and I was still needing a lot of oxygen. They assessed my movement and tested my reflexes. My left leg was totally unresponsive, as was my right arm. I don't remember being especially concerned but I think I was still really confused about it all. This was a different team of physios so I wasn't familiar with them and I didn't really fully understand why they were there.

STEPH

It was unbelievable when the physio rang me and put the phone to Jen's ear. It was a one sided conversation as she couldn't speak but the physio said she saw Jen react and smile from something I said. The physio was really lovely and obviously very caring of Jen. Hard to imagine this, but she told me that Jen had been riding a bike above her bed and she had cycled half a mile! Of course the pedals were moving her feet but how wonderful that was to hear. I don't think Jen was too impressed knowing that she was most likely undressed at this point but I'm sure they were very respectful.

A few days into January, things were really beginning to happen. Jen was gradually having more time off the ventilator, there was some eye contact and they said that she was trying to speak. They told me that a speech box would be fitted to the tracheostomy. I didn't really understand what that was but her voice was still very weak and quiet. I can't imagine what she must have been feeling. Everything must have been so scary and strange for her. I really wished I could be there to help and reassure her!

By the 9th she was managing to be off the ventilator during the day and was speaking very quietly but the doctor said we need to remain cautious.

A few days later, I had a call from the hospital. My heart jumped every time that number came up! I expected to hear the doctor's voice but unbelievably it was Jen speaking. Her voice was breathless and extremely quiet but she was determined and tried so hard to speak to me. The physio was holding the phone to her ear. I'm not really too sure how aware Jen was but it was a lovely surprise for me – although quite an emotional experience.

Dare we now start to believe that she was coming back to us?

Well after a couple more days she was taken right off the ventilator and had the tracheostomy removed. The doctors were amazed at her sudden progress and still advised caution but suddenly she was beginning to use her phone with lots of help from the physio and she even sent a message to a friend. I think she might have sent some messages without realizing as they were slightly garbled and didn't make much sense but she was aware enough to try.

On Friday 14 January I got a call from the hospital to say that at last Jen had tested negative for Covid and I could go in to see her that afternoon. What a shock that was! As I was just leaving home a doctor rang with a warning to be prepared to be meeting a very different Jen from the one I was used to. That journey to the hospital was difficult. What did he mean? What had happened to her after everything she had been through? I didn't know what to expect and felt quite nervous.

Covid rules meant that visitors were still not allowed but they had pulled out all the stops for us as they thought it would help her recovery. Two nurses brought her down on her bed to a special garden room where we spent an hour together. Obviously the nurses had to stay close and monitor how she was doing throughout. I am so very grateful to the nursing staff for organising that visit. Jen was a bit confused and found it rather difficult to speak, her voice was still very soft and breathy – she was obviously struggling to produce sound. It was so good to see her after so long although it felt very strange, I didn't know what to do or say. Could I give her a cuddle or should I keep my distance? Poor Jen did get upset and the nurses reassured me that a hug was very much needed. Despite the doctor's warning, she was very much still my Jen!

Her hands were very shaky and I was surprised to see how thin and bony they were. Her beautiful hair was long and quite untidy, her eyebrows had grown so much – I knew she would hate that when she saw herself! That hour seemed to go so fast. The nurses were lovely – they kept an eye on Jen but at the same time gave us privacy. When time was up the nurses took a photo of us together which I couldn't wait to share with everyone when I got home. It was a very emotional visit she was tearful and trying to make sense of what had happened and realising just how much she had missed whilst 'asleep'. It was obviously very confusing for her and I don't think she remembered much about the visit.

I am so very grateful for being allowed that visit. It was so kind of the ward staff to organise it. Apparently it's very unusual for an ICU patient to be allowed a garden visit. (I was later told that the 'Peace Garden' is there for terminally ill patients to meet family)

By the next day Jen was texting and making calls and on Sunday the 16th she asked me to bring cakes in for the nurses as they were talking about moving her to HDU. She was doing so well.

12

ON THE MOVE

As I became more lucid I was aware that I was actually now moving to another ward. People were packing my things up – and there seemed to be a lot of things! They told me I was moving to HDU. I was really frightened to leave ICU – which makes no sense as it was such an intense place, but it was safe. There was someone there all the time and, other than the one time that was probably a dream, I hadn't been alone at all. They told me it was a positive step and that I didn't need the one to one care anymore but that there would still be people looking after me and that I would be able to get help easily if I needed it.

They told me about how they had been very worried about me at times because it was looking like the outcome wouldn't be positive and many doctors were calling me a miracle. They actually came just to see me because they couldn't believe I was conscious and communicating. I didn't feel much like a miracle at all. I seemed to be well known on the ward and everyone who came in appeared to know me. I think I had been there for so long and been through so much in there that a lot of people had taken care of me. It was like being a celebrity – but not for particularly good reasons! One nurse told me she had been assigned to me one night and I was incredibly poorly. She told me she wasn't expecting me to live through the night and was surprised but relieved to see me still there next time she came in.

She said she didn't take her eyes off me the whole time she was there with me. A long night shift!

All my belongings were piled on my bed and I was told that Mum was outside the ward. She had come to bring me some things and they had asked her to wait as she could see me when I moved between wards. For some reason getting out of ICU then seemed to take ages and I was really worried about Mum standing outside. She had hurt her leg and I must have been aware of that because I can remember asking repeatedly if she had a chair out there. I didn't know what was outside of the ward – whether there was a waiting area or if anyone would have at least helped her to find a seat.

It was well over an hour before I was ready to leave the ward. As they pushed me out on the bed I saw other patients. I'd had no idea they were there! They were in a ward next to each other whereas I was in an area by myself surrounded by machines. I'm almost certain I wasn't surrounded by other people each side. I was really shocked to see them there and tried to work out whether they were in intensive care or a different area. They appeared to be conscious and unventilated, however my perception wasn't particularly strong so I could be totally wrong!

Anyway, I was rushed along the corridor and I saw Mum but I didn't get to stop and see her properly. The whole way to HDU I kept telling them that she wouldn't be able to keep up with them and that she had arthritis. I kept asking them to check whether she was okay but we forged ahead regardless.

We arrived on HDU and I think it must have been early evening. They told me I would have my own room and when they wheeled me in, it was a beautiful room. It had big windows. Looking out across the town and it was sunset so it was bright orange. I don't actually know if this was real or not!

When I got there I asked about Mum and eventually they

told me she couldn't come into the ward and I couldn't see her. Covid regulations. I remember crying because I knew she had been waiting and would have been hurrying to keep up with us. I wanted so badly to see her. I felt terribly guilty for making her suffer and unbearably sad that I hadn't been able to see her more than fleetingly. What I needed more than anything at that point was a hug from her and it was denied. It felt so unfair and so difficult to comprehend. Hopeless. Knowing that she would have just had to wander on home having not seen me anyway. I had also left an environment that was familiar to me, one that had become a safe place where everyone knew me. I was now in a room by myself. No noise, no regular turning or one to one care. No beeping. Just alone trapped in a still partially confused and delirious mind.

I was however elated to see that I was in my own room with my own bathroom. It wasn't until they all left me that I laid in the bed and wondered how I was going to actually get to the toilet. Ever hopeful, I formulated a plan in my head that I would speak to the physio when she came and tell her that I needed her to get me up so that I could just walk to the toilet. That's all. To the toilet. Of course when it came to it, it wasn't that simple because my body wasn't strong enough to do any of that.

The lead physio, Bea came in and got me up to a seated position on the bed and I can remember being really pleased that I was sitting without being held but I then suddenly couldn't catch my breath at all and it was quickly very frightening. There may have been an element of panic in there but I physically could not breathe and she very quickly laid me back down on the bed. She told me afterwards that my oxygen levels dropped rapidly and my lips started to go blue. She said it was actually quite scary for her too! I suppose stroke physios don't always have to deal with post Covid complications in addition to the plethora of stroke and ICU related symptoms. She was excellent and made sure I was fully okay. Despite this episode, I still had no idea that I wouldn't be able to walk!

Once I was settled, Susan, a speech and language therapist, came in to assess my eating and drinking again and she bought some food and fluid samples for me to try so that she could see how I handled the different textures with my healing tracheostomy incision. I was so determined to show her I could swallow perfectly well that I probably became a little defiant. There was no way I was going to let her tell me I couldn't swallow well enough! Admittedly I didn't find swallowing easy. I was terrified of choking and I could feel that it was about to happen but I pretended I was certain I was fine and I did manage to swallow.

Susan decided that I could start eating soft food and said that they would take my order for supper and I could order something from the soft menu. I chose macaroni cheese and peas. Ironic really that this would be my first warm proper food as it's my favourite meal! I was really looking forward to it! Sadly the message didn't fully get through and what I actually got was puréed macaroni cheese and puréed peas. They brought it in and when I said I was allowed soft food now they apologised but said it was too late to change it. I gave it a go. Oh my goodness, it was so tasty after only being allowed 100mls of thickened fluid or puréed pear for what felt like forever! I needed some help to feed myself which I didn't like much but I did try to do some of it. I realised I could only move my left arm and couldn't feel or move the right. I wasn't sure why that was but I was thankfully able to use my left, even though it was very weak. I wasn't able to eat much of the macaroni cheese but, as first meals go it was very nice!

I hadn't been able to use a bedpan or soil myself on purpose since being a little more lucid so I literally hadn't emptied my bowels at all for a few days. I wasn't really eating much but I knew I needed to go. So, once I was on my own, I returned to my plan. I was going to ask the physio Bea to just get me so that I could sit on the toilet – I didn't care how I got there. I knew I couldn't feel my left leg but it didn't really occur to me that at least transferring from one seat to another wouldn't be possible. I think I knew by now that I wasn't going to just get up and walk but I thought I must be able to at least slide onto the toilet. The relief of that thought was huge. The torment of the toilet was almost over and I was going to get to sit on a proper toilet for the first time. Thank goodness I didn't know how long it would actually take!'

Sadly my plan didn't come to fruition again. After Bea had sat me up she came back and I discussed my plan with her. I didn't understand why she wasn't particularly keen to get me up and straight to the toilet. She didn't say I couldn't but kindly tried to steer me away from it, suggesting the bedpan may be safer – it had been a long day with the move from ICU and I was exhausted. So I accepted that I'd have to wait, assuming I would go the next day. The nurses that evening said the same and offered me a bedpan a few times. I refused.

The next morning I was allowed a Weetabix with milk – on the strict condition that I mushed it up. I told them that was fine and actually how I would have it anyway so that was good. I enjoyed eating that – very slowly! I also enjoyed my first unthickened coffee! It all tasted so good!

After breakfast three nurses came in and asked if I was ready for my wash. I knew I wasn't going to like it – I also knew I had been washed already countless times but this time I was a lot more aware – the drugs I had been on had obviously been drastically reduced by this point. It was one male nurse and two females. The male was guiding the females but not actually washing me himself. So, in my head, he was essentially just watching me, naked and being washed in the most intimate of places and, unlike in ICU (or at least from what I remembered), they didn't cover any part of me to do it. I am not particularly comfortable with being naked around other people at the best of times but I felt really vulnerable at this point and it was really quite upsetting. I eventually asked them whether they could please cover me with a towel which they did. I was very pleased when they had finished!

Next the two assistant stroke physios, Dominic and Tyson came in to see me – both lovely but very young and male. I was in a gown but it wasn't done up so, again I felt quite vulnerable although they were both very respectful. They got me sat up and the aim was to be upright for a little longer than the day before. It did involve them holding my mostly naked body in awkward places and I was very conscious of that. As soon as I was upright I, again, couldn't catch my breath and although I endured it for as long as I could, it wasn't long. I asked them to lay me back down which they did but they lifted my legs onto the bed at the same time as laying my head back and I swung very quickly down into laying flat. At which point the room began to spin. I have had a lot of dizziness in the past with neurological issues but have never experienced dizziness like this. It was extreme. I felt like I was going to pass out – actually I felt like I was going to die. Dominic stayed with his hand on my shoulder until it eventually stopped. He said maybe we took the movement too quickly and once it had started to ease off, he slowly tilted the bed so that I was propped up a little. It passed and I was okay again. It was really scary.

Once Dominic had gone I desperately wanted to speak to someone from my family just to tell them – another thing we don't realise now is that our first thoughts are often about who we can tell about things that happen. We video and photograph the events in our lives – not only to capture memories, but so that we can post updates to a large number of people by clicking a few buttons. Now that you don't need to be face to face with the person you want to tell and you can literally talk to people all over the world just by posting on social media, I think we forget how things were before technology was as advanced as it is now. But I still wasn't really able to communicate to my fingers how to use my bloody phone. In real life we spend so much time using devices and it is second nature but there was a complete block between my mind and my fingers and I just couldn't work it out. I had no way of contacting or speaking to anyone and I started to feel a bit hopeless.

In reality, I had gone from one to one care to being alone in a room. I had a call button but if I didn't call (which I didn't), no one came. I hated to be a bother so I just sat on my own staring into space which I appeared to have gotten very good at! I wasn't thinking about the future much yet because I was still trying to get my head around the present.

13

18 JANUARY 2022 – STEPPED DOWN TO STEP UP

The ward doctor came in to see me and suddenly they believed I was well enough to go to the ward. I'd only spent one night on HDU and I was then being packed up again to move to the ward. It was okay though because they said that they were actually moving me to another private room on a ward. I knew I still needed to use the toilet and was relieved that I would still have my own bathroom. Plus I am much happier not having to share a bed space!

Once I was packed the nurse announced that actually there was a bed for me on the stroke ward. It was a general ward – not the private room they had said about. I started to panic. It might sound overly dramatic but I had spent many weeks either asleep or in delirium, been moved for one night to a room where I suddenly needed to be more independent – i.e. feeding myself actual food, and I was now being moved again to a completely new environment with other patients and the toilet became a massive issue once again because there was just a shared bathroom. I asked how many people were on the ward and was told around twenty.

By this point I was really panicking and started to cry. I was picturing an old fashioned ward with lines of patients all in one room. Nothing was familiar. I couldn't see my family and the place I thought I needed to get used to was changed very quickly. It was completely overwhelming. I was so desperately distressed that I think I actually got cross with the nurses who had told me I was going to my own room because they had given me the wrong information and it was a huge deal to me. It was everything. All consuming and so far removed from anything remotely familiar.

In all the hospital stays I have had I've only been on a ward once as luckily I've had my own room. It was actually okay that one time but in this moment the prospect was terrifying. I was dependent on staff to do everything for me and very quickly had to adapt to new people and places, noises, smells and routines. If you've never been in that position I can tell you it feels like the world is ending. How much more can one person take?

But I needed to move and, aside from not having any choice, people kept on telling me it was great that I was well enough to step down onto a ward. Of course it was but I still wasn't fully aware of what was going on and was pretty much in shock from learning what had happened to me. So a change to the type of environment that is my worst nightmare was frankly too much for me. I felt really guilty for making a fuss and getting cross with people but it felt like a massive leap into the unknown to me at that moment and it was very frightening. The only slight comfort was that the physio team was going to be the same neuro physio team I had seen in ICU and HDU so Bea, Dominic and Tyson would be there. I clung onto that tiny bit of familiarity.

My belongings were piled onto my bed again and I was wheeled off to the stroke ward – still sobbing. My things were put into a cupboard (which I was unable to reach) and there I was. I was reeling for quite a while and I think I had to sit and think about what had just happened for a long time.

Eventually I became too desperate for the toilet and knew

I had to do something. I was adamant that a bedpan was out of the question. I should just explain that after surgery when I was around eighteen I was put on a bedpan whilst laying flat. This just does not work with the female anatomy – in fact any anatomy, and my bed had to be changed afterwards – but I was left in that wet bed for an entire night first. Thankfully I still had a catheter this time so that part was taken care of but my mental block towards the bedpan meant that I couldn't and wouldn't contemplate it. Even in a state of delirium and confusion I was stubborn as hell.

Eventually I was so desperate I agreed to try the commode as a compromise and used it successfully by being hoisted onto it. The whole time I was aware of the other people in the ward behind a thin curtain. It was mortifying but I had no option because it had been days and I was in pain. Thankfully, although the ward did have around twenty patients as I was told, each bay only had four so it wasn't as terrible as I had been imagining. There was no privacy though. Everyone could hear everything through the curtains.

I was hoisted back to bed and I think I went to sleep. It was noisier on the ward and sadly, being a stroke ward, the majority of patients were elderly and very poorly. Some were awake during the day and some were asleep all the time but there wasn't any conversation – I probably wasn't really very communicative anyway.

At around 9pm on my first night on the stroke ward, a doctor I'd never seen before came in and said he had been called in to carry out some nerve conduction studies to try and find out why I couldn't feel my right arm or left leg. I was really exhausted but he had apparently driven over from London to do this study – and was driving back afterwards. I spent a while wondering why on earth he had done that just to see me! Maybe I misunderstood

– I was quite confused!

He stuck stickers on me and sent electrical currents into my limbs. Then he stuck needles into the muscles of my arm and leg. I was just totally dumbfounded that I couldn't feel anything! Even though I knew before the test that I couldn't. I watched the needles go in and you automatically wait for the pain but there wasn't any.

After a while he said I had critical care polyneuropathy in both limbs caused by being in ICU. I asked him if the feeling would come back and he said he thought it may well improve but it would most likely be slow if it returned at all. He also said I had carpal tunnel syndrome in my right arm which would be contributing to the numbness and would likely improve with surgery. Then he drew out some charts, gave them to the nurse and vanished.

Nothing was done electronically by him and those tests inevitably got lost. Much later a nurse asked me if I thought I had imagined the whole thing. I know I wasn't fully myself but I don't think I did imagine it. How would I know about the needles? And I'd never heard of critical care polyneuropathy. I'd heard of carpal tunnel syndrome but I didn't have that before so I am pretty certain it happened. Whether it happened exactly as I remember I don't know but it did happen!

The next morning, Bea came to get me and told me we were going to the gym. I wasn't too sure what to make of that but they rolled me from side to side until a sling was underneath me and then hoisted me into a chair. They wheeled me into the gym where three more physios were waiting. Bea explained that we were going to try standing. I was so relieved thinking that soon I'd be able to get myself around a bit better.

They surrounded me and told me to try and stand. I did what I have done since I was around one year old and tensed my leg muscles to lift myself up. Nothing happened. That was a bit of a shock! Four young strong physios were holding me, pulling me or pushing me using various parts of my body or the gown I was wearing (and yes, it had the delightful open back!) and I couldn't make any movement to stand. We tried a few times until I just couldn't do anymore. I was completely exhausted. They were very encouraging and told me I had done really well, then hoisted me into the chair, wheeled me back and hoisted me into bed. I didn't feel like I had done well at all!

It was at that exact point that I started to think about what was ahead of me. I felt like I was at the bottom of a very large mountain and I had absolutely no idea what life was going to look like. Was I going to be in a wheelchair for the rest of my life? Would I even get out of hospital? Would I need carers? Once I got back to my bed I was alone once again mulling over what was ahead and it was pretty terrifying. A total game changer. Before Covid I was working, driving, walking about and generally getting around independently. That seemed so far away from reality at this point.

I had lost everything.

I spent one night there before they came in and told me I was moving to a different part of the ward. Immediately. They packed all my stuff up and wheeled me to a different bay with my belongings. This was another four person bay and again, the other three patients were elderly. I was apparently moving because each section of the ward I moved to was less intensive than the last and it was good because I was getting better. It didn't feel good. It just felt confusing and disorientating to suddenly be somewhere different again. I was upset and crying again. My emotions were all over the place.

I spent one night there before the same thing happened again. I found myself being wheeled along to a different bay.

New people, new surroundings, new patients – still elderly and largely unresponsive most of the time. I was very unsettled and churned up. I felt like I was being shoved from pillar to post and I still don't think I was completely with it. I stayed very emotional and tearful.

The ward occupational therapist Sasha came to see me. She was lovely. She was so kind and understanding. I told her about being moved everyday and she was horrified. She went to speak to the staff on duty straight away and told them that under no circumstances must I be moved again. She was concerned that I was a stroke patient with some confusion and the constant changes in environment just added to my confusion. She then came back and reassured me that I was there to stay and wouldn't move again. Patients were often wheeled away with no notice and I thought it really can't be just me that struggled with that! I was actually then moved once more but only to the bay opposite as it was bigger and I was being hoisted which took up room.

As I started to tell the story of what I knew about being unwell so far to Sasha, her shock was a total surprise to me. I still didn't even begin to have any real understanding of why I was there or what was happening to me. I was telling the story that I had been told. I had, and still have, very little memory of it. Her surprise weirdly reassured me. She told me that yes I was facing a mountain but she would argue that I had already climbed the hardest part just by surviving. I hadn't thought of it like that at all. I asked her if she thought I would walk again. She said she didn't know but after everything I had fought she thought it was likely I'd try my hardest to do that.

It was the strangest feeling to suddenly have people telling me about how hard I had fought and being nice to me as doctors in my life have often seemed to trivialize things and so many have been unkind. Focusing (rightly so) on the bigger, more unwell cases and mostly ignoring me – but I was still suffering even if it wasn't quite bad enough. Maybe I've been unlucky with some doctors I've seen in the past. Suddenly I was that more unwell case! It was a really strange feeling because, although it was horrible, confusing, frustrating etc, it was also a bit of a relief that I was accepted as a valid patient.

If there was one thing I could ask of the medical profession it would be to never dismiss someone telling you there is something wrong. It doesn't mean it has to be dealt with in any more urgency than anyone else but, despite what you might think that patient is feeling, no one ever knows what another person is feeling. It doesn't matter how many cases of migraines you see, there will be some people who can put up with the worst pain and carry on as normal and some who are just in unbearable pain and floored by it. You just don't know. If it's not life threatening, it doesn't mean it isn't life destroying. Sometimes there is more going on than meets the eye – as we discovered when I was initially diagnosed with lupus after suffering years of symptoms which were largely dismissed and not joined together.

My most developed and fine tuned skill is my front. Sense of humour, self deprecating, upbeat (mostly). A lot of the time it is a very well sculpted and practiced shell and it doesn't often change unless I am really struggling or am asked the right questions – or I'm at home. It isn't done on purpose and it certainly isn't always a good thing but it is the way I am. It's the way I have become after too many bad experiences. Protect myself and don't make a fuss.

I believe that the best and most effective doctors are those who take the time to ask the questions and break through the shell. The pain becomes obvious if you can just crack the front. And when you do crack it, it can be life changing for that patient. Had my lupus been diagnosed earlier it may have alleviated a lot of physical suffering and mental anguish. The patient can never truly diagnose themselves – we simply don't have the tools or the knowledge, even with the internet. We have no option but to rely on the medical profession and when we are telling them there is something going on there most probably is. Just because the most obvious thing can be discounted doesn't mean that person's pain is not there or doesn't matter.

Anyway, Sasha began working with me on trying to get some function back in my arm. She suggested I massage it with different textures and also that I start trying to use it. It was a strange feeling being unable to feel it and having to try to move it. First she wanted me to try holding a knife with my right hand. I couldn't even make it move towards the knife but after trying consistently I managed to start moving it. It had no strength so picking up a knife was really difficult – especially as I couldn't feel my middle three fingers and half of my thumb but I persisted. I think the small amount of feeling I did have in specific parts of my arm helped me to start moving it. Over the next few weeks she continued to work with me and help me to get it to move and gain a little strength. I did jigsaw puzzles and activities that we generally use in the preschool at work just to try and reconnect my brain and my fingers. Slowly I managed to start using it although it was fairly limited and very weak.

My right arm remained numb and continues to be so. This is now thought to be nerve damage from being proned in ICU. I have been told that this happens because people have to be put in unusual positions to help the lungs and quite often this causes compression on the brachial plexus nerves which can become damaged and this, combined with the ulcer, appears to have been the cause of the numbness which will now most likely not improve.

The physio team on the other hand, were working on my

legs. Every day Bea hoisted me out of the bed into a chair and wheeled me around to the gym. To start with, I was just trying to move my legs. Waking up my hips and knees. She would put something shiny under my foot to make it slippery and I had to try to slide my feet backwards and forwards along the floor one at a time. My right leg co-operated and moved, although it was very weak. My left leg did not cooperate at all at this stage. All I could do was look at it in disbelief and will it to move.

Bea then put a cycling machine in front of me. I was a bit surprised given that I couldn't move one of my legs but she explained that the machine had a motor which would help me. She also told me that I had cycled half a mile on a similar machine while I was on the ventilator! And apparently laughed while I was doing it – I must have been high! They said the bike was suspended above me. The mind boggles – particularly now I consider that I was dressed only in a gown with no underwear! Don't picture it! So anyway, back to the gym, she strapped my feet onto the pedals and off we went. I have an early photograph of myself doing this and am still shocked by how skinny my arms and legs were.

I was still being bed bathed. The care assistants would come around in the mornings and work from one patient to the next. I dreaded my turn as I hated it so much. I was more able to communicate now though and, given my bedpan/commode fear, we started to get a routine where they would hoist me out of bed onto a commode in the morning and then wheel me straight into the bathroom where they were able to position the commode over the toilet so that I could, at last, use it! It was a relief in more ways than one! They would then pull me around to the sink and help me to wash. Whilst I was still having help to do the washing and unfortunately to clean me after using the toilet, it felt a little more private and normal being in the bathroom.

I was able to see Mum one more time once I was settled on the ward. The staff arranged for me to go to the garden room again. This time I went in a chair. It was a 'tilt in space' chair so was reclining and supportive. But it was dreadfully uncomfortable because I couldn't move my body. As I was in a chair and had been off the ventilator for a little while I was taken down by a porter. Mum came in and then we were left alone. She was able to hug me and I needed it so badly. Time went by so quickly and before I knew it the porter was there to take me back and we had to say goodbye again. We had no idea for how long.

STEPH

By the time of Jen's move onto a general ward (with two more moves over the next couple of days) she was able to use her phone. There were lots of frustrated calls and tears. This was a stroke ward with all the other patients being rather old and sleeping most of the time. In a short time she had woken up from the coma, started to realise just what had happened to her and was trying to get her body moving. Luckily she is extremely stubborn and worked hard to get as mobile as she could. One arm and one leg were out of action and she was very shaky, breathless and still needing oxygen. But considering how she very nearly didn't make it, she was actually doing really well.

However, the phone calls got more and more difficult as she went through so many horrible experiences. I desperately wanted to be there for her and reassure her, I think it would have helped if I had been able to see where she was and meet the people caring for her. I was working in the dark and could only imagine what it must be like. There seemed nothing I could do to help her. There was still no visiting and if I rang the ward it was not at all easy to get any information.

In hindsight now I think I should have insisted on being able to talk to a doctor. But I knew just how busy everyone was and I didn't want to make a nuisance of myself. Jen is an adult and once she was awake it seemed that I should take a step back. The staff spoke to her but didn't expect to tell me things. Looking back now, I realise that Jen was not understanding what had happened, what she had gone through, and was very muddled between the delirium experience, confusion and coming to terms with the way her body was now. She needed an advocate.

When I saw her for the second time she was in a chair and didn't have the accompanying nurses or in fact any medical staff at all. I was so relieved that she was obviously so much more independent, even though much of her body was not yet working. She had her own pyjamas on and looked a lot more like my daughter. What a transformation in such a short amount of days.

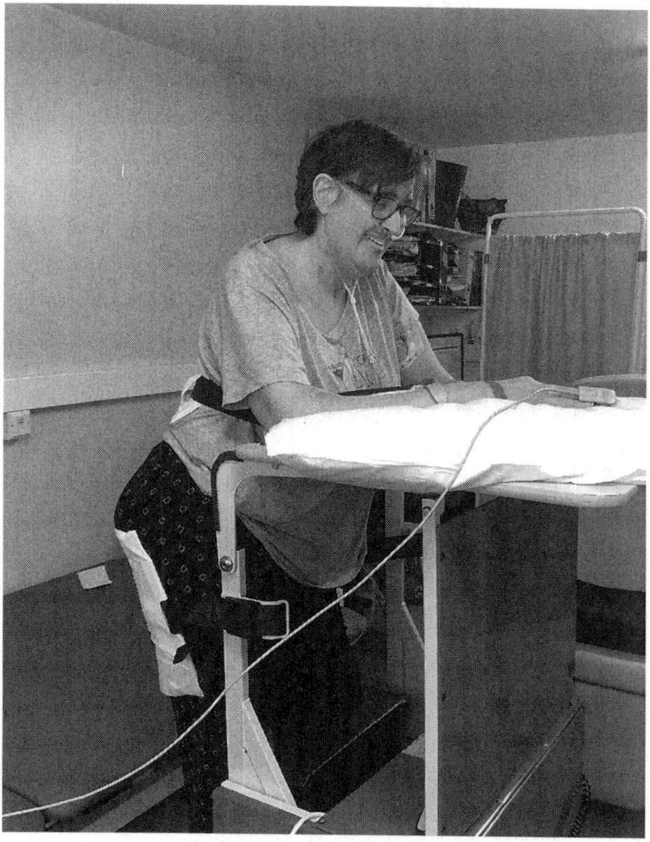

Learning to stand

14

STARING UP AT A MOUNTAIN

A few days in, one of the care assistants asked me if I wanted to have a shower. I couldn't believe it! I was very grateful. The bathrooms hadn't been fitted completely correctly and water would leak under the door when someone used the shower so she had to put a horseshoe of towels around me on the floor before we got started and then two carers showered me. I had to grit my teeth as they touched areas that I would rather have washed myself but it was worth it as I actually felt fresh and clean! It was such a good feeling after so long.

Washing in the bathroom and using the toilet wasn't without its pitfalls as it meant I usually had to wait until everyone else on the ward was washed and often the bathroom (which was shared by eight beds in a 'Jack and Jill' style – serving two four bedded wards with a door on each side) was occupied when it was time for me to use it and I had to wait.

I still had a catheter but was needing to wear a 'pad' (an adult nappy that isn't allowed to be called a nappy) for emptying bowels. Once I got into a routine of being hoisted onto the commode and wheeled to the toilet I didn't really need the pad at all. It was so uncomfortable and one morning it was stuck in an awkward place. I didn't have the strength to move any part of my body to dislodge it and became really distressed. Being so uncomfortable but completely unable to do anything about it at all. Communicating was difficult still and it's very hard to ask someone just to 'please pull the nappy out of my bum'!

Bea, who was fortunately outside my curtain when this happened, came in and quietly suggested that I might prefer to wear my own underwear. I couldn't believe this hadn't even occurred to me. So I stopped wearing the pads and was able to wear my own underwear. To get my knickers on though, I had to be laid on the bed and had to pull myself over onto one side then roll over to the other side. I then had to do this all again so the carers could pull them up for me one side at a time. I didn't have the strength to turn myself so, even though I did try to help, I still had to be rolled mostly by the carers. Still pretty undignified but better than the alternative.

As I got stronger, Sasha and the OT team encouraged me to start trying to wash myself. We started with just my face, then my upper body. It was such hard work! Even just trying to hold the flannel was really difficult but I persevered as much as possible.

The ulcer on my right arm was being dressed every few days and the nurses were very concerned about it. It had developed a large black top and was very deep! It was insanely painful and I was put on strong painkillers to help. The edges were rolled over rather than flat so it was very difficult to see how it was going to heal. It certainly wasn't getting any smaller.

Bea suggested I might like to try and put some pyjamas on and I was really keen to do that so I decided I wanted to try for a shower by myself – I thought it might be easier than trying to reach different parts of my body with the flannel. So the carers set me up and left me to it – at my request. It was so much more difficult than I expected! I was sat on a shower chair but had to try and hold the shower head up to wash myself as well as apply soap. My left arm was not strong enough and my right was pretty much useless so it turned out to be a very quick shower!

I dried myself as best as I could and then tried to get my bra on. I never even left my bed without a bra on previously and I hadn't had one on at all since I went to sleep. In the gown it didn't seem to matter much but it did in pyjamas. I tried really hard to put it on myself but I couldn't get it behind me, I couldn't untwist it or pull it around myself. And that was before I even attempted the clips! I had to abandon it before I even got anywhere near to it being on. I was so frustrated with myself! I managed to get my t-shirt on but had no chance of getting my pyjama bottoms on by myself so I had to pull the help cord and wait for help to come.

The help cord appeared to be not working. That was unfortunate. I was pulling it, half naked and completely stuck in the bathroom by myself. I waited for a while but no one came. I was trying not to panic and was looking around at the other emergency cords around me. I couldn't reach any of them. Couldn't move myself anywhere. It was a horrible feeling because I had no idea how long I might be there. I waited a little longer but no one came so I had to bang on the door and hope that someone heard me. I had barely any strength to make a loud knock and I could only just reach. Eventually someone did hear me knocking and came in. She was an OT visiting a different patient but she said she would get me some help which she did. I'll never forget that feeling of complete helplessness and hopelessness.

When the nurses came back in, I was pushed back to my cubicle on the shower chair and hoisted (with no pyjama bottoms on) back onto my bed. Once on there, my lower half needed to be dried and then I had to roll from side to side (being pushed and pulled) while they inched my pyjama bottoms up. Was it worth it? Yes, totally! I felt a lot more comfortable than

I had been in the gown and, even though I found it difficult, I had showered!

Next Bea had decided that I needed to start sitting in the chair for increasing periods of time. It was another super comfy looking chair with tilting and reclining so I was looking forward to sitting in it. To get into it I then had to be hoisted again from my bed and into the chair where I was left with my iPad in front of me. I should have been quite comfortable but within about half an hour, like before, it was the most uncomfortable I think I've ever been. What we don't realise when we sit is that we are constantly moving even if we don't notice. Shifting from one side to another, crossing and uncrossing legs and generally staying comfortable. I couldn't do that.

The ward staff would try to encourage me to stay in the chair for increasing amounts of time until I was in there for a couple of hours. But then they would tell me to just stay until after lunch which I did. Then 'after lunch' came along and they were so busy feeding other patients, cleaning them up and having lunch breaks themselves that I was often left in the chair for a lot longer. I can remember sobbing and begging them to get me back to bed but it needed two staff to hoist and I kept being told to wait until they'd finished their lunch breaks. Awful for them as they couldn't help me but to me it was actually like being tortured. I was sat on a chair right next to my bed. I could touch it but I just couldn't get onto it! I was helpless and totally dependent. It was even worse than I could ever imagine.

I decided that I needed to get out of the chair and back to the bed before lunch in future to avoid being stuck in the chair for longer. You'd imagine that sitting in the chair would bring relief. Even now I can't imagine being just in bed all the time but back then I would have given anything to stay in my bed and just be left alone.

The chair bought nothing but pain and discomfort. I couldn't even watch the iPad to distract myself because I wasn't able to concentrate on anything except how I was going to get back to bed. But on the odd occasion that I didn't sit in the chair (weekends), I felt extremely guilty! I knew that I couldn't live the rest of my life in bed, I just couldn't take the pain of sitting and couldn't see how that was ever going to get better. Luckily, as I regained some movement I learnt to sit for longer but it took weeks.

When Bea and the physio team came to get me they were able to wheel me in the chair round to the gym which saved another hoist. Once I was there they hoisted me out and put me on the couch sat up. We were going to have another go at standing. I wasn't keen because it had been soul destroying the first time but Bea said she had a machine that could help me.

It looked like a torture instrument and I couldn't believe such a thing even existed when I saw it. It looked like it had been made in someone's shed in their garden and I thought they were joking. It was wooden with a tray (like a baby's highchair) and a number of straps. Both my legs were strapped to it and the hammock I was sat on was hooked to it. I then had another strap tied around my back. They pressed a button and it slowly started to rise up. Both physios then had to push/pull me up and with the force of all of that I was upright. My arms were on the tray and I was 'stood'. There was very little weight on my legs and my left leg wouldn't even straighten. By the time I got up there I was unable to breathe and got to the point where I started to panic so they sat me down again. I was 'stood' for about ten seconds. It took a long time (or what seemed a long time) for me to recover and when they checked my oxygen sats they were down in the low eighties – I'd had oxygen on the whole time with nasal prongs so we had to increase the intensity and wait for my breathing to slow down. It was exhausting! They put me back in the chair and wheeled me back to my cubicle where they hoisted me straight back into bed because I was tired and sitting up hurt so much.

When I laid there and reflected, I was still trying to make sense of it all. I hadn't been able to wash, dry or dress myself but here I was having managed to have a shower (of sorts) and I was dressed. I'd got up onto my feet! Even though it took a motor to get me there and I couldn't stay there, I was up there! Could I really come back from this?

I was back on a full diet and ordered every meal but to begin with I just couldn't face food at all. I felt a bit sick and I struggled with breathlessness when I tried to eat anything. Who could have imagined that eating would cause breathing difficulties! I was stopping the oxygen most of the time but still needed it for physio and it was suggested that I kept it for eating too so I had to put nose prongs in for every mealtime. I also needed help to cut up my food as my right arm still wasn't strong enough to do anything and my left arm wasn't especially useful either. I did generally manage to feed myself though when I felt able to eat.

The dietician was involved because I was still being tube fed. They generally did the feed overnight so I wasn't particularly aware of it – other than the tube up my nose. She was worried that I wasn't eating though and made me also have Fortijuice, which is basically fluid with nutrients to boost the body and give more concentrated calories. I hated it and avoided drinking it whenever I could. I couldn't believe a dietician was telling me I must consume more calories! Never in my life did I see that coming! I kept trying to think of ways to get rid of the bottle without drinking it. Difficult when you can't do anything for yourself so often I just drank it like a shot to get it out of the way.

I had started to notice that I was having lupus rashes on my face. I would be sitting still and could feel the rash creeping up my neck and down my nose. It feels like a feather tickling to start with and then the intensity ramps up to it being quite sore. People started to comment that I looked flushed. I knew what it was. I wasn't taking my lupus medication because I had been so unwell with the sepsis and a flare was coming as a result. It started with the rash which got much worse than it had ever been before. It very quickly spread to my whole upper body and was burning. I asked to see the doctor who took one look at it and called the rheumatologist. He had seen me in ICU and had been monitoring my lupus whilst I had been very unwell. I took some photos of the rash just in case it went before he got there but it just so happened that he arrived as it was in full swing. He agreed there and then to put me back on the immunosuppressants to treat the lupus and took some more bloods. He couldn't understand it because he had run all the lupus antibody bloods in ICU and they had been negative – despite being consistently positive when I was diagnosed. The bloods he took were still all negative which was very strange and initially he said I may have been 'cured' by Covid but evidently that wasn't the case! We agreed to continue the medication regardless of the blood results.

I was starting to suffer mentally and of course it was understandable given the situation but I had noticed that I wasn't on my antidepressants. I'm guessing they were also stopped when I was very unwell. So I asked to restart them because I could feel a fast decline and I was worried having been on them for years previously. The particular drug I was on was unlicensed in the UK so the doctor said he would ask the consultant. Nothing much happened. A day or two later a lady came into my cubicle and she looked nice. I felt desperate and begged her to get my meds restarted – I had absolutely no idea who she was – she could have been anyone! But it turned out by chance that she was actually the consultant, Dr Lark.

The meds couldn't just be restarted because of the lack of UK license so Dr Lark said she would speak to my psychiatrist about how best to reintroduce them. She tried to call him and he tried to call her back but, given that he was close to retiring and only worked one evening a week, they didn't appear to be able to catch each other. I think I asked her every day whether she'd got in touch with him until his secretary emailed to let me know he had sent a letter in the post! It felt like weeks but was probably a few short days before they did finally manage to speak and the medication was reintroduced.

The ward psychologist, Laura spent a lot of time talking to me and trying to help me start to make sense of it all. She was Canadian and was lovely. I found her easy to interact with and she seemed to identify with how I was feeling. She also knew of my medication as it is used in the US fairly frequently. I was starting to remember things from ICU and they were confusing. I remembered about the police calling and I couldn't work out if it happened or if I dreamt it. It had felt so real to me that I couldn't understand the possibility that it wasn't real. So she kindly went up to ICU to ask. At least I think she did – she may well have realised that there was no way it actually happened and not checked but, either way, she reassured me that I could chalk that particular event up as a definite dream. It helped to know for sure and to have someone who could check these things for me. She helped me a lot whilst I was on the ward. I don't actually recall her talking to any other patients. Of course she must have done.

15

LIFE ON A STROKE WARD

By now, I was having physio daily with Bea and OT input from Sasha most days. Both teams were excellent and seemed to work together to look at the whole patient. On one morning Bea, a different OT and Laura all turned up at the same time. Unfortunately I was having a bad morning. I had patiently waited to be hoisted to the toilet and to wash as I generally waited until everyone else was finished – they all needed to be washed in bed so I didn't want the nurses to have to mess about with hoisting me while they were waiting. In actual fact it would have made more sense for me to go first so I could wash and dress in the bathroom while they were with the others but it didn't work out that way. Anyway, on this morning I needed the toilet and had asked to be taken in sooner rather than later. I waited an hour and then asked again because I was getting desperate but they were so busy and other patients (from the next bay) were in and out of the bathroom. I was just sat on my bed starting to feel unwell.

So when the OT, Bea and Laura arrived, I was past the point of desperate and had had an accident (I still had a catheter so it was the dreaded bowels again). I was devastated. They had come to see me all together to look at helping me holistically but I was a quivering wreck who needed to be changed and cleaned up. Laura left whilst Bea and the OT hoisted me out of bed and into the bathroom. Bea then left and the OT stayed to help me get cleaned up and dressed. She was very kind and sensitive. She said it was a good opportunity for her to observe me washing myself and dressing as much as I could (my top half). I won't ever forget the feeling of being absolutely desperate for the toilet and being unable to do anything at all about it. It was nobody's fault but is totally dehumanizing.

We'd been slowly increasing the amount of time I spent standing in the mechanical frame. I had built up from a few seconds to thirty seconds, a minute, two minutes. It took an awful lot of gritted teeth and stubbornness! I needed oxygen to do it and really struggled to catch my breath every time. It became a really scary experience and I genuinely thought I wouldn't ever get on my feet by myself. But slowly I did manage longer times. Bea would set her phone timer and put it in front of me so I could watch the time ticking away. Somehow that helped.

Laura did come with me to physio on a different occasion so that she could observe my breathlessness. I think they thought it might be anxiety and in many ways it was probably added to by that but after observing me she said she thought it was more physical and that we should try to think about some techniques we could use to stay calm and control my breathing to increase my lung efficiency as there was clear lung damage. Bea talked about square breathing where you find a square to look at and breathe in as you move your eyes up one side, out when you move your eyes across the top, in when they go down on the third side and so on. I tried really hard but I struggled to do it. They really wanted me to breathe in through my nose and out through my mouth but when I tried it I literally felt like I was suffocating. Of course it was terrifying.

Eventually we moved on from the equipment that mechanically pulled me up to stand, to a standing frame where I had to get myself up but as soon as I was up, a seat folded down underneath me. It would help me on the ward rather than having to be hoisted because, once I was perched on the seat, I could be wheeled around. It was called a Sara Stedy.

It usually took me a few goes to get myself to the standing position using the Sara Stedy and I needed at least two people pulling/pushing me up. It was so hard. My muscles were terribly weak in my arms and my legs and neither had the strength to get me up. I felt like I was tensing my thigh muscles as hard as possible but they were unresponsive. But the physios were very strong, luckily! It was worth it to stop needing the hoist – which was never comfortable, so I stuck at it. But getting to a standing position led to many comical and inventive ways of people pulling me – usually involving pulling me up using my clothing – not so easy we found when I was standing from using the toilet with no clothes on my lower half to grab. I was pushed and pulled in a variety of ways that I don't care to remember to get me standing and plenty of times I thought I wouldn't get up at all! The bed was fairly easy as it was height adjustable so we made it higher but the toilet was a fixed height and was much lower – as was the dreaded chair. Can you imagine being sat on the toilet half naked thinking that you just won't be able to get off? It was exhausting. But it was slow progress.

I was unusual on the ward in that I was a lot younger than the other patients – being a stroke ward. The three other ladies in my bay were, for the longest time, very elderly and often sleeping. Unfortunately that brings many of its own challenges. They generally needed feeding and changing regularly. They needed washing and dressing and they didn't often get out of bed. They were obviously very poorly. Generally people would be moved out of the bay and instantly replaced with someone else. Some, I suspect, were moved and passed away pretty quickly afterwards.

A new lady arrived who was a little younger and we spoke a bit through the day. She hadn't actually had a stroke but had been unwell and was put in that ward as that's where there was a bed. She was friendly enough and it was nice to have someone else in there who was awake and communicative.

However.

As soon as the lights went out she turned into the actual devil. I've no idea what happened but she started shouting and asking for the light to be turned on. The nurses were trying so hard to keep her quiet but she just got more and more aggressive. There was no way I was going to get any sleep. It literally went on for hours. I felt so bad for the nurses because they were being given a really tough time and for her because something had obviously happened and she must have been really frightened. The staff were so patient.

More senior staff were called eventually and at one point there must have been about eight people in her bed space. I heard them trying to get her to take something to calm her down but she wasn't having it. She was demanding things and at one point, the patient next to her – who was asleep 99% of the time piped up with, "Please would be nice." Now this patient never spoke and didn't seem to really know what was going on most of the time but she was clearly a very intelligent and polite lady who never complained or argued unless she was in pain. It made me chuckle when she suddenly spoke up with that! And the unsettled patient did then say please. Briefly.

In the end they sedated her forcibly. It was awful to hear. I really wanted to go to sleep and shut it out so badly but it was just so disruptive. I even tried putting on my noise cancelling headphones but I couldn't sleep with them on. I could hear her saying things like "You're not injecting me", "No! Stay away".

It was distressing. Eventually I asked for a sleeping pill. She had been sedated so I fell asleep. It must have been about 3am. At 6am she started again. I could have cried – I suspect the nurses felt the same! Anyway, we were all got up as usual, washed etc and she then started talking to me completely coherently again!

She was moved later on that day. It turned out she had a UTI which I know often makes older people behave unusually. I hope she recovered well.

She was replaced with a ninety-four year old lady who was completely deaf. I had very little voice still after the ventilator and trache so it was really difficult but we did manage to converse a little with hand gestures. I was exhausted so wasn't particularly sociable and I'm fairly sure I asked to have my curtains pulled round – not to be rude but just to try and have a bit of time on my own. I joked with the nurses that we were having a quiet night that night because I was so tired.

However …!

The lady had a catheter and a stoma and she obviously did not like either so as soon as the lights went out she started moaning and then shouting for help. She wanted them to empty the stoma bag which they did. But she then wanted them to do it again, then empty her catheter and this went on for hours.

Being totally deaf she may not have realised that she was shouting at the top of her voice. The nurses were, again, trying to keep her quiet but couldn't – and of course they couldn't talk to her quietly because she couldn't hear them. So at 2am I asked for another sleeping pill. I was in total disbelief that this was happening for a second night in a row! The next day this lady was also moved. The nurse told her she was being moved because it was a stroke ward and she hadn't had a stroke. She pointed at me and said, "What about her, has she had one then?"

I was tired and annoyed by then and said "Yes I actually

have", and she moved onto another ward. I so badly wanted that bed to stay empty!

It didn't. A few hours later another lady arrived. This lady was a little younger and had a head injury that caused a bleed rather than an age-related stroke. I was praying that she would sleep at night! We started talking and, although she obviously felt very unwell, she was a nice lady – and she was quiet!

By then I was so tired that I did spend quite a lot of time with my curtains pulled around because I just didn't want to see anyone. I had been around people with no escape for a couple of months by now and I was tired of interacting. I sometimes liked to have some time alone in normal life and wasn't one to be chatting with other people all day so it wasn't unusual to want to be on my own but I'm sure it looked like I was just incredibly depressed. I mean I wasn't mentally on top form but I was really just craving some quiet time.

I did see the team in my little 'cave' though. As well as the regular eye drops I needed, there was the meds, painkillers and bloods that were taken every day because I was taking warfarin – all of which involved people visiting me in there.

I still had the large ulcer on the inside of my elbow and everyone who saw it gasped because it was horrible. The dermatologist had been a few times and, given the sepsis and lupus, was reluctant to put me on the strong immunosuppressant that Pyoderma Gangrenosum needs so he opted to treat it conservatively with daily dressing changes, cleaning and cream. There was talk of a skin graft by the rheumatologist but the dermatologist just said that my body was eating itself and would just eat the graft too so there was no point going through with surgery when I wasn't really well enough only for the pyoderma to destroy the graft. I genuinely didn't believe there was any way it could heal itself but, as predicted by the dermatologist, it did eventually.

The other doctors on the ward weren't quite so encouraging. They came on their ward rounds each morning but it seemed to be different ones all the time. They would come in and ask me if I needed anything. If I did, they would half heartedly examine whatever it was and then tell me nothing that I couldn't have worked out myself! They would also invariably tell me to drink more water. I would then hear them visit every other patient in my area of the ward and say the exact same thing. I asked one to look at my neck once because I had a red lump and didn't know what it was. They said it was just a spot so I assumed it would go away. It didn't go away and I then figured out it was actually a scar from the central lines I had in during the coma! After that I tended to listen more to the consultant when she came round with them.

16

FAIRY LIGHTS AND MEMORIES

Bea was absolutely determined that I was going to walk. She decided to try electrical stimulation to try and wake up my left leg. She put electrodes on my leg and started with the electrical current, slowly turning it up. I couldn't feel it at all and there was no movement in my muscles. She couldn't understand it and kept on trying but there was no response. I asked her if I was likely to walk again and she said she just didn't know.

She came back a few times, refusing to believe that nothing was responding to the electrical current. She decided to try it on my good leg. No response! So maybe it was just me. Or perhaps I was just too weak generally.

Weekdays on the stroke ward were fairly busy with the various specialist teams and nurses/assistants bustling around. Weekends, however, were horrible. There were still no visitors allowed and at weekends there was a skeleton staff. Just enough to be 'safe'. No physio and no OT. I dreaded every weekend. The staff were so stretched that it was difficult to get their attention when I needed anything so I really just kept myself quiet in the corner. Some weekends I washed and then went back to bed and stayed there. There were no physios to insist I sat in the chair and it was easier for the nurses not to have to hoist me in and out of bed so they were happy for me to stay there. I was quite happy not to have to sit in the torture chair. Probably wasn't the best – and certainly wasn't good for mental health but that's just the way it was.

The lady opposite was still the younger lady who had the fall and we chatted a bit. She was very quiet and was obviously feeling very unwell. She spent most of her time in bed – even when the physios came and got her to sit in the chair, she got up and went straight back to bed when they left. I was so jealous as I was stuck in the chair! I don't think anyone actually realised how just how poorly she was. She also got very visibly depressed with being stuck where she was with no family to visit. I could identify!

The lady diagonally was older and had obviously had a stroke. She wasn't really able to do anything and slept a lot but when she was lucid she was animated. It must have been very difficult being stuck in a body that wouldn't do anything and a brain that wasn't functioning quite as it should. I so badly hope she recovered some sort of life but sadly I don't expect she did. There was certainly no improvement during the time we were on the ward together.

I couldn't understand why I was made to get out of bed every day and no one else in the bay seemed to be. I suppose it was sadly because I was younger and most likely to recover. Or because I'd arrived at the stroke ward with slightly different conditions. But in my head it was just unfair. I watched everyone snoozing in their bed just desperate to get into mine. I know it was best for me but it didn't feel that way.

As we weren't able to see anyone, the nurses did get family to video call patients and they would bring the phone in so they could see their family – obviously some didn't have, or couldn't work, their own mobile phones so there was a ward phone.

They tried their hardest to bring familiarity in but sometimes it was actually more distressing watching them see people they loved and trying to communicate with them. It must have been so hard for their families. I cried a few times listening to them – especially when it was obvious that the family members were desperately trying to get their mother or grandmother back and weren't getting the responses they were hoping for.

I was lucky to have my devices and by now I was able to work them. I video called my best friend Lynsey quite early on. I just remember her being in the car (not driving luckily) and bursting into tears when I first called and she didn't stop the whole time we were talking. I found it difficult to understand what on earth was wrong with her! Obviously, along with my family, she had struggled during the intensive care period. It was so distressing for everyone. I felt really guilty that I had caused so much pain and suffering. It was really good to see her and have a link to life before. We were talking (well I tried) for quite a while before I got a bit tired. We vowed we would talk more and, not only have we done that but we are closer than ever. Covid didn't and wouldn't ever break us.

Lynsey sent me a big box of gifts while I was on the ward. And it really was a big box! It was full to the brim with drinks, a few snacks, colouring books and pencils – and fairy lights to decorate my cubicle. Sasha put the lights up for me above my bed. She told me that infection control would most likely ask me to take them down. They never did so they brightened up my cubicle until I went to rehab – where they did the same thing. Lynsey probably could never know how much that parcel meant and the simple thought of fairy lights made a big difference – not just to me as it cheered up the staff too!

By now I had a little collection of photographs that Mum had brought in whilst I was in ICU. They had been put up on the whiteboard in my cubicle. Once I was on the ward the nurses slowly started to put them up on the wall until I had a huge photo and card wall. It was so lovely to lay in bed and look at it whenever I was feeling a bit low. With the fairy lights and photos, it was fairly easy to see that I'd been there for a while!

STEPH

On 2nd February – hooray, rules were relaxed so visits were finally allowed, one nominated person for one hour but negative tests had to be shown on the way in each time. Oh my goodness I always did the test with some trepidation in case I got a positive but luckily I never did.

Over the next few days Jen became unsettled and emotional – she even said she wished she hadn't woken up. So hard for a mum to hear that but I fully understood why she felt that way. Life had changed in a big way for her, a completely new way of living had to be learned. In fact I think our whole family was struggling to come to terms with all that had happened – we all wanted to help her as best as we could.

We started thinking how the house could be adapted and made safe for when she finally came home. It was a bit scary and this whole experience made me realise that we really would need downstairs living for her. Also I began to realise the need to live closer to my son and his partner. Moving house had been thought of many times over the past few years since my husband passed, but I had lived in the same area for over fifty years and Jen all her life so each time we got close to moving we didn't manage to make it happen. Such a big step! So many people to leave behind! But this was the right time, this was necessary now, we need to be closer together as a family. I am getting older and Jen is now disabled. I'm sure friends and neighbours thought we were taking a ridiculous step by moving. With all that Jen had been through, we didn't even know how able she would be yet, or when she would come home. But in my heart I knew that at last the time was right to put the house up for sale.

17

HYDRO-MOTION

I continued to work on trying to stand. Progress was slow and it was really tough. It honestly felt like I was never going to do it at all. Bea came to see me one day and said she was thinking hydrotherapy might be good. She explained that standing and walking with the support of the water would be much easier. I had told her before that I enjoyed swimming and she obviously remembered. I didn't realise there was even a hydrotherapy pool at the hospital! She said they hadn't done it before from the ward so she would need some time to organise it. In the meantime she thought I would need to have watertight covers on both arms – my right because of the ulcer and dressing and then the PICC line (a semi-permanent cannula) which was still in my left arm where medication went in and blood could be taken out for testing. When it was looking like it could happen she ordered some arm covers but they were really slow to come so I went online and bought them myself. They arrived the next day and Mum was able to bring them in. I also bought some shorts and a vest top as it would have been really difficult to get a swimming costume on.

I was really nervous when it was finally time to do it. I was dressed in my shorts and vest top and they hoisted me onto a special shower chair which they wheeled around to the hydro pool. They were really fussy in there and the wheels of the chair had to be sterilized before we were allowed to even wheel in there. Bea was there in her swimsuit with another physio and they both got into the water. My attractive arm covers were put on both arms up to my armpits and a third physio wheeled me to the pool where her and a fourth physio hoisted me up in a special pool hoist. They lowered me into the water and the feeling was amazing! The water was warm and soothing and I felt really calm.

I held onto a large float that went around my whole body and with a physio either side of me I was up! I couldn't believe I had managed to stand! After all that time and with a lot less assistance than on dry land. They held the float and slowly we started to move across the pool. My legs were moving and I was 'walking'! My left leg, which I still couldn't feel, kept floating up so the physio on that side kept standing on my foot to keep it down! I was able to make the walking movement and it was so good. I was very breathless and needed oxygen the whole time so one of the physios on dry land was looking after the oxygen tank. We were probably only in there for fifteen minutes but I was utterly exhausted!

But the downside was having to get dressed on the other side. Once we had finished I was hoisted back out into the shower chair. We decided that, rather than hoist me onto a bed in the hydro pool area, dress me, hoist me into a chair and then hoist me back into bed, it was easier to wheel me back (still wet) to my cubicle – it really wasn't far. So I was covered in towels and wheeled back. I was allowed to go straight back onto the bed as I was shattered.

Once I was back in my cubicle I was hoisted onto the bed which was covered in towels and we started the mammoth task of drying and changing me. It was another 'grit my teeth' moment as the physios set about getting me dry by rolling from side to side on the bed and then getting me into my pyjamas in the same way – and then changing the wet sheets. Put a bit of a dampener on the experience but I was so happy that I had been able to actually do something with my body!

I was really keen to do it again and Bea said she would try to organise it. They were very particular on the balance of chemicals in the pool (they obviously use it for a lot of people and Covid was still rife). On my second planned visit the chemicals weren't right and I wasn't able to go which was disappointing – I was dressed in my shorts and t-shirt ready and then had to change back into my pyjamas. I did manage one more session in the pool before I went to rehab and then a few sessions as an outpatient after I went home.

I also started to attend a few upper limb group sessions. Every time they came to get me, I'd had physio already so I was knackered in bed but, since I vowed to myself that I wouldn't say no to physios and OT's, I went along. We did a variety of exercises in a group setting – there were only around four people. They'd all had strokes obviously and had various levels of impairment. The sessions really helped me to get some movement back in my arm which was still mostly numb and very weak.

There were some truly wonderful staff on the ward. I really don't know how they do the job they do. They have a lot of very poorly patients and most are elderly. Most need help with everything. There were outstanding staff members for whom I will forever be grateful. Bex who made me laugh every time she was in – even though she made me stay in the chair for longer and actually ran away when I asked her to get me out early! Then there was a student nurse. I am so sad that I can't remember her name. She was lovely. Took time to really care for everyone. She spent time talking to the patient diagonally from me who was largely unresponsive but when she sat and took the time to converse with her, she came to life. I didn't see it with anyone else. She was being listened to. A few times I saw people feeding her and she was trying to tell them something but they needed to get the job done and move on. I wished I could get myself over to her and just listen to her but I couldn't.

On one day I went down to physio in the morning. Bea wasn't in so I was with someone else. We did some exercises and were sitting looking at other exercises I could do back on the ward when I saw two people walk in from outside. They had two dogs with them! My dogs!

I was utterly dumbfounded. I had no idea they were coming but the physio team and Mum had arranged it all. The room was suddenly full of people. The dogs, Ivy and Poppy, were excited to be somewhere different but they didn't recognise me at first – my voice sounded very different. But slowly they started sniffing me and it was very obvious when they realised it was me! They were so happy, licking me and wagging their tails. It was such a special moment. I had missed them so much and it was amazing having them there. It really cheered me up. They didn't stay for too long and I was taken back onto the ward but seeing Mum and the girls filled me with a little bit of energy to keep going.

STEPH

Jen was working incredibly hard to get some mobility back. She learnt to write with her left hand as the right one had been affected by nerve damage. But during this time Jen did become very low, frustrated and unhappy at times. It was difficult for her to understand just what had happened to her. She started to have disturbing flashbacks which were very distressing.

To try to help and motivate her, the very kind physios arranged for me to bring the dogs in to visit. Gosh it took a lot of organising, but what a surprise she got when we all trooped in during a physio session. I think all the physios were just as excited as she was during that visit! The dogs were certainly pleased to see her as well and they loved all the attention they got from everyone there. The physio team were absolutely amazing. I know that she would never have got so far without them, they even organised for Jen to have a couple of sessions in the hydrotherapy pool. As she was still on oxygen she needed four physios each time but these were amazing sessions for her.

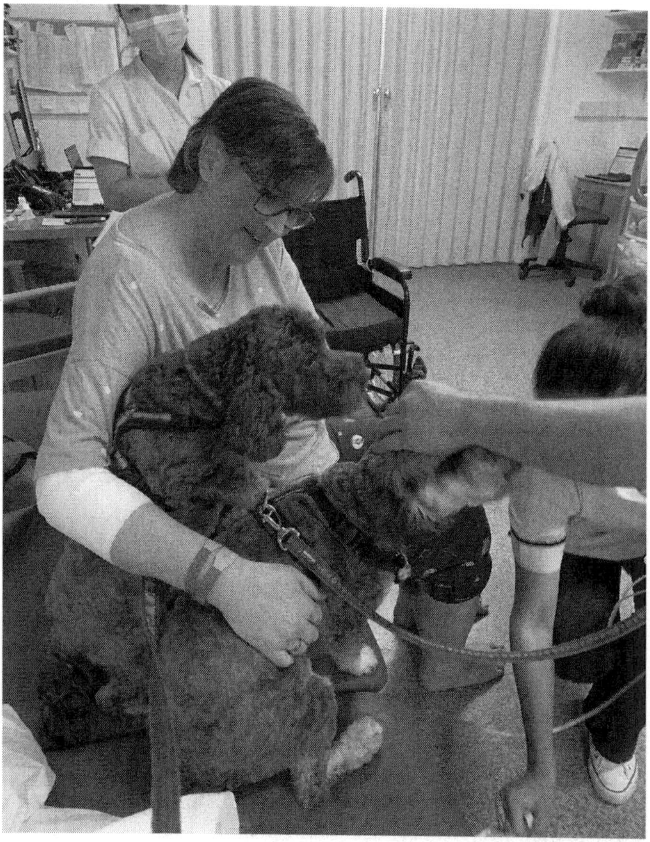

A visit from Ivy and Poppy

18

LIFE ON WHEELS

There were some very dark times on the ward and I shed a lot of tears. Everything was so overwhelming and I was still trying to process it all. But I was also trying to work out what life was going to look like. It was very difficult to see past the ward. To imagine anything other than being hoisted everywhere and being unable to do anything for myself. I kept asking what my recovery was going to look like and people were positive but noncommittal. I recognised and understood that.

One day Dr Lark came to visit in the morning as normal and as she left, she told me she would come back to have a chat with me about the future.

She came back later and basically told me to prepare myself for being a wheelchair user full time. She felt that, whilst my breathing and leg problems may improve, they weren't going to improve vastly and in her opinion I should start thinking about how I could manage that at home.

Although I had bad moments, I think I generally accepted everything pretty well and when we had this conversation I think I just said okay, thank you. I'll start thinking about it. It was actually refreshing to be given an insight into what life may look like after hours and hours by myself just wondering. She set the bar pretty low and that was what I needed. I needed to know. Anything above that low bar was a bonus. I understand why the team were noncommittal – they truly didn't know what my life was going to look like and didn't want to promise anything in case I didn't achieve it. For me personally, I needed to know the worst case. Because anything over and above that is success. If someone tells you that you will be back to normal and then you don't achieve that level of progress it feels like failing. I discovered this later on as people made assumptions about how much I could progress and insisted on giving me the best case scenario. It's definitely great to be positive but feeling like a failure when you don't achieve that level of recovery is emotionally crippling because you feel like you've let everyone down. I appreciated this consultant being honest and realistic based on what she could see.

Eventually, as the Covid situation continued to evolve, visitors were allowed back. We were able to have one person and only that one person could visit for an hours prebooked slot each evening – and only one visitor in each bay at any one time which meant fitting in around the visitors for other patients. So finally Mum was able to come in and see me. It was lovely to see her again and she helped me to sort out my cupboard. Cupboards in hospital appear to be completely useless because they are situated next to the bed but by the wall so that once the bed is sat up, the cupboard is miles away. I couldn't reach it at all. Couldn't see my pyjamas to ask for specific ones and I couldn't control how my clothes were put back into the cupboard. I found that very frustrating! I wasn't able to sit myself up or swing my legs over to the other side of the bed to check the cupboard so I was stuck wherever I was left.

I felt so guilty that Mum had to drive all the way over when she was obviously having trouble walking. I knew she would feel like she needed to come every night and I also knew there was no way that was feasible! So she came every two to three days

– even that was too much for her really. The lady opposite had her husband in every night and I remember being a bit jealous. But they lived closer so it was easier. It was a long way for Mum to come for just an hour.

I had collected quite a lot of things by this point as I had been in hospital for weeks and Mum had been bringing things I needed into me. When she wasn't allowed to visit she would drive over and leave a bag in reception. Can you imagine knowing your loved one is inside and very vulnerable but being unable to see them? I can't imagine how she must have felt. The nurses commented a few times about how much stuff I had – it just accumulated!

I had a wheelchair at home so one day I asked Sasha if mum should bring it in. I had bought it years before just to go to Canada. I'd never used it otherwise. She thought it was a good idea so I asked Mum to bring it in for me. She told me that she had actually hurt herself and wouldn't be able to bring it in herself but maybe Andrew could bring it. We were still only allowed one person to visit but the nurses very kindly bent the rules and let me see him outside. So I put on warm clothes and was pushed outside in the wheelchair to see him. It was freezing but it was so lovely to see him – we weren't allowed near each other. I don't know what he thought because I looked terrible with my hair all over the place, unwashed and with huge un-plucked eyebrows! We talked for a little while but I did get tired very quickly and was taken back to the ward. I felt really guilty because they had made a special concession in letting me see Andrew and the patient opposite me had asked to see her son but wasn't allowed. I think I had just been there for so long that the staff felt I needed a bit of a pick me up. I really did.

The wheelchair came to physio with me and I was hoisted into it. Sasha and Bea felt it was too small so Sasha found me a hospital one. I was very excited thinking I could move myself around in it but it felt like a tank and I couldn't roll myself further than the end of the bed before I was exhausted! I was still just too weak and my right arm especially couldn't make the wheels move. I was so disappointed.

I started to notice that there was hair appearing on my pillow, my clothes and in the sink when I attempted to wash it. I started to become aware that my hair was definitely getting thinner. It had always been really thick but not anymore. It was lanky and thin. My scalp was starting to show and I began to think about just shaving it off. I wasn't bothered about having hair particularly – it was a pain to wash with only one arm and I didn't have access to a mirror at all so the only idea I had about how I looked was using my phone and whenever I used zoom or FaceTime. The consultant suggested early menopause so ran the relevant blood tests but they all came back negative. The only explanation was stress and severe illness.

As no one other than Mum could visit there were quite a few Zoom and FaceTime calls. Lynsey called me most days which I loved and I had family Zoom calls with mum, Andrew and Phil together. It wasn't quite the same as having someone there just to talk to normally, to have physical contact with another person who wasn't doing something medical to you! I loved the zoom calls but they did make me feel extremely homesick. I was always a little downhearted after them. Being with the people I loved seemed so far away at a time when it was so important. I wore headphones but was very aware that everyone could hear me talking even if they couldn't hear the people I was talking to.

I even zoomed into a choir rehearsal! It was so amazing to see them all and they had all been so worried. When I first called my friends Judy and Lynne from choir they were both crying. I still couldn't understand why. I wasn't that important was I? I always thought I was a bit of a pain. A burden ... The rehearsal was wonderful. They sang me a song and it meant so much to me as choir had been an enormous part of my life.

I spent a lot of time thinking about my homecoming. In my head it was idyllic. Everyone was crying and pleased to see me and we had a huge party to celebrate with all the family and my friends there – all in Andrew's garden (I hadn't consulted him on this). They would, of course, be desperate to see me and it would be so memorable. Well it didn't turn out that way – realistically it was never going to. I still had no idea how ill I had been and if I had, I would have realised that I wasn't going to be all better when I eventually did go home. But I had missed Christmas and I just wanted to recreate it when I got home – despite it being summer. When it came to it, it just wasn't feasible and I really wasn't well enough.

19

MOVING ON AND TO THE SIDE

About two months after going onto the stroke ward we started talking about what was going to happen when I was discharged. I was told that I needed to go to rehab – there wasn't a choice in that. They thought I would probably spend around three months there at least.

I was given a leaflet on the rehab place. It looked quite good with lots of physio and some mental health help too. I was pretty fed up with being around elderly women and I was really desperate for some time on my own. I was starting to leave my curtains closed more and more because I just didn't feel like making conversation. I asked if I would have my own room in rehab and they told me it was either single or two bedded rooms. It sounded better than where I was but I didn't like the thought of sharing with one other person. You have to constantly be with that person and I like my own space. At least with four people you can choose not to converse if you want to but it's a bit more awkward with two. But there was no choice so I started to try and get my head around it.

One morning Sasha came in and mentioned that actually I might go to rehab somewhere else. This had never been mentioned before and I did freak out a bit. "How is Mum going to get there?" – I knew it was way too far for her to come regularly. But I had a PICC line in my left arm and still had the open wound on my right which needed dressing every other day. I also still needed oxygen whenever I moved and for eating. She explained that I couldn't go to the original place with those three things. I felt like I had just got my head around one place and then another came to light that I never even knew was an option in the first place. Change was unsettling and, again, I had no control.

As I was still taking warfarin I needed daily bloods to check my INR (the clotting potential of my blood) and they used that measurement to give me the correct dosage of medication each night. Dr Lark told me that they were thinking about removing my PICC line as, apart from the warfarin monitoring, I didn't need it and it would be best to go to rehab with as few infection risks as possible. So she suggested that we stop the warfarin and try something different. I was a little hesitant as I know that clotting caused by lupus is usually treated with warfarin and other blood thinners are not so effective but she thought it would be better. I had also tried a range of alternatives when I first went on blood thinners and they all made me feels very nauseous. I tried the new drug she suggested and was fine on it so the decision was made to remove the PICC line. This PICC line had provided a semi-permanent solution to giving IV drugs as well as testing blood so it had definitely been my friend in lessening the number of other needles I had to endure. The nurse came to remove it and took all the sticky dressings off. Then she pulled it out. It was so long! Maybe my imagination or the dream side of my confused mind has remembered this in an exaggerated way but it seriously looked long enough to navigate around my entire body! I was shocked!

One night it suddenly occurred to me that I had been sleeping on my back. I never used to sleep on my back – I was a front sleeper. I knew I would probably have been snoring on my back but I wasn't physically able to turn myself. I was turned by staff but invariably that just involved tilting me a different way and putting a pillow or blanket under me to prevent pressure sores.

There was a male nurse on duty for the night and I decided to ask him to help me turn onto my side. It's pretty unimaginable to suddenly be unable to do even these simple things – even now remembering it still feels like I am telling a story about someone else. I just could not move my body other than hauling it from side to side (with help) to get dressed and when I did that my upper body moved but my legs didn't follow at all. Anyway, he managed to turn me fairly easily onto my side and that night the relief was amazing! I actually felt comfortable!

Of course my lack of movement meant that I was in that position for the whole night and when I woke up in the morning I couldn't get myself off my side when I wanted to be on my back to sit the bed up for breakfast. That was frustrating and I spent a few mornings twisted to sitting whilst partially still essentially on my side. I didn't want to be a nuisance so endured it until I was hoisted out to get washed.

After that night I slept on my side every night and the staff had various techniques of helping me over. I was able to pull myself using my left arm and the cot side but my legs still had to be lifted over. Some preferred to have me pull on the side bars whilst they pushed me over. That was the most difficult. There was the one nurse who grabbed and pulled the sheet quickly from under me – now I think about it, it was a bit like those people who pull the tablecloths out from under the crockery! I stayed in place but just popped over onto my side with enough sheet beneath me. It worked a treat! And then there was the nurse who basically lifted me up by my pyjamas and twisted me over – it was quite a fine art! They all then needed to position my legs so that I stayed on my side.

Even though I couldn't feel my lower left leg my foot kept on touching the end of the bed and it was complete agony. It must have irritated whatever nerve was sleeping enough to send it crazy. But I couldn't move it myself so if it got stuck I just needed help and the pain was so excruciating that I needed it quickly. Of course it isn't that easy in a large busy ward so often I was crying by the time someone was able to help me and just move it away from the end of the bed. The pain was like nothing else! I dreaded it happening! It seemed like such an innocent thing with a simple solution but it was so painful and I couldn't do anything about it myself.

A couple of months after I had woken up, the team called a family meeting to discuss me moving on to rehab. Nobody actually told Mum and she got a call about forty minutes before it was starting and I was just wheeled in with no warning. Mum made it to the meeting but Andrew had to zoom in. The occupational therapist explained that they felt I was well enough to move onto rehab and they indicated that I would be going to the alternative rehab with the oxygen I still needed being the main reason. I was really against going there but Andrew told me he had looked it up and it looked okay. He reassured me that it was single rooms only. That pretty much sold it to me and I felt a bit better about it.

The staff knew by now that I needed warning about any changes so, once a bed in rehab became available, they gave me a few days' notice for the move. I felt really frightened about being so much further from home and from mum. My mind went into overdrive imagining what it would be like in rehab. Would they be strict? Would they make me get dressed every day? Did I need mum to bring me clothes as all I had was pyjamas?

I was still toying with shaving my head and asked if Mum could do it at my bedside or in the bathroom. The staff said she could do it anywhere. She came in but we decided that maybe she could just trim it with scissors which she did. We had heard that there might be a hairdresser in the new hospital and I thought maybe I could ask them to just shave it off. Or even cut it to hide the bald patches.

I still had a catheter and wasn't getting any sensation when I needed to pass water. I was changed onto a flip/flow valve instead of a bag which meant my bladder filled up rather than the bag and I was able to empty it myself – in readiness for doing that naturally. Essentially, my bladder wasn't getting the opportunity to feel full when there was no resistance and it was draining straight to the bag so it didn't need to wake up. The flip/flow was basically a tap that was put on the catheter and then when I felt I needed to go, I was able to empty it myself and to start with I just did it every few hours. I still wasn't feeling anything. The nurses kept telling me to give it time but I wasn't convinced I'd ever feel it.

I discovered that the catheter emptied most effectively when I was actually sitting on the loo – so when it was pointing downwards. As I couldn't get to the toilet by myself or dangle my legs over the side of the bed, I started emptying it by moving my bed so that the foot end of my bed was pointing downwards. I can remember Sasha walking in and giggling at me as I was sitting really high up on the bed with the foot end down and a bottle between my legs emptying it. I wanted to make sure it emptied properly so I could get away without doing it for a while and I couldn't sit up by myself to do it! I still had the flip/flow in when I went to rehab.

In my last week a new patient came in who was next to me as the existing patient was moved to an area of the ward with more care. This lady was ninety-four and the first thing she told me was that she was a bit deaf. This immediately filled me with bad memories of the other lady but this lady decided we were going to chat no matter what. Even though I was into my 'leave me alone' phase she would potter up to my bed for a chat every now and then.

Pat was a tonic. Our conversations were random but I learned all about her son (who she took great pains to tell me was single and my age). Here I was stuck in my bed in my pyjamas, being hoisted to use the toilet, with a catheter and thinning unwashed hair as well as monster eyebrows that hadn't seen a pair of tweezers in months. I probably felt about as unattractive as it was possible to feel so I just laughed along with her. But she did make me laugh and I enjoyed her visits to my bedside. She was very sprightly, very independent and I was so incredibly grateful for her.

Pat was only there for about three nights – and I did meet her son! It turned out she hadn't had a stroke so went off home.

Pat wrote to me a few times after she left – even before I came out of rehab. She obviously had a collection of postcards and she would send me a selection with some handwritten notes. It was as eccentric as she was but made me smile every single time!

I later heard from her son that she had in fact had a fairly large stroke a few months later but she was cognitively okay and happily I got an email from her afterwards and a card at Christmas – even two years on. She may be a little older but she's obviously still Pat and she's fabulous. I'll remember Pat very fondly forever. She was a shining light in my really dark world at that time.

My physio was progressing and, determined that they would see me walk, I pushed myself really hard to stand (with the help of Bea and another physio) and take a few steps with a walking frame. I could barely breathe and it was harder than you can imagine but I really wanted them to see me take a step before I left because they had worked so hard with me considering where I started. They were obviously really pleased.

STEPH

On 21st February Jen was finally moved to a rehab hospital, it was further away than we would have liked but she was still needing oxygen. She also had the nasty ulcer on her right arm that needed to be dressed daily.

Once again visiting was not allowed – a bitter blow for Jen. This was a time when she needed her family more than ever! It also proved to be even harder for me to get to speak with any of the doctors or nursing staff. In this hospital, patients were in rooms on their own so unfortunately she became quite isolated, we spoke on the phone and did have FaceTimes together but that was often very difficult.

Through much of this time Jen was unhappy, we had some difficult phone conversations. I have since learned that her care seemed rather inconsistent. Through her calls she told me some of the things that happened and was often in tears. But I could never get to speak to anyone. I felt like a nuisance! I also kept telling myself that they knew what they were doing – they must know best. Sometimes you have to be 'cruel to be kind' don't you? I didn't however understand just exactly what Jen was going through at that time. Looking back how I wish I could have been there with her.

It was a stroke unit and, in hindsight, it was not the right place for her to be. I don't feel the staff there fully understood exactly what Jen had been through since that fateful 1st November when Covid struck.

It has taken a long time for Jen to be able to understand and talk through the things that happened to her during her stay in rehab and I have been able to listen to her, look back and now understand the reason for the many upsetting calls to home. She had to deal with all the repercussions from the delirium she experienced, the changes to her life, having to use a wheelchair and all that it involves.

Jen has since been persuaded by many people to write down her experiences and this she has done. By doing that I think it has helped her to understand, to make some sense of what has happened and try to start a new life. A different one. I am not going to dwell on this time in rehab as this is detailed in Jen's story.

I do remember her trying to tell me things that happened at the time, but not being there with her I didn't really take in what was happening. I found it so upsetting to see it written down once she finished this account but difficult as it was it was definitely a good thing to do. It has helped her to gradually come to terms with what happened. We hope that maybe it will also help others (staff and patients) some day.

20

SLIDE BOARDING

I was moved to rehab in late February 2022 by ambulance. Mum was allowed in to help me pack as I was panicking that the ambulance would only take one bag and I had a lot more than that! Sounds silly now but at the time it was a source of great anxiety. She went through everything and helped me to pack what I needed. She said she would drive over to rehab with more of my things. Sadly visitors weren't allowed at all over there so having just been allowed to start seeing Mum I then couldn't see anyone again. It was a bit of a blow.

The ambulance eventually arrived to take me and everyone waved me goodbye – I had been there for a few weeks and, on reflection, I was getting better where I suspect most of their patients don't so it was probably nice to see me leave doing better than when I arrived. I had built relationships with a lot of the staff who had cared for me and it felt like a huge wrench moving on. I was extremely anxious about it. I stifled yet more tears.

I said goodbye to Mum and went off to rehab. I didn't know what to expect – I didn't even really know where it was! On the entire journey I was worrying about Mum having to drive on the motorway or on big busy roundabouts. The ambulance technician with me was chatting away trying to distract me but I kept asking where she was and panicking that the roads we were on were big roads and she would be scared.

We arrived after about half an hour. It seemed a lot longer! The hospital looked fairly modern and was more like a community hospital rather than the big hospital I had come from. I was taken straight into my own room and was shown my bathroom. It was shared with the room next door which was slightly disappointing but it was certainly more private than where I had come from and I was craving privacy.

The nurse checking me in was called Nula. She was quite brusque and took my details. I then remember being examined all over my body looking for marks or pressure sores. I'm pretty sure this happened but my memory isn't totally trustworthy! I was looked over with no clothes on and every part of my body was checked. I suspect this is protocol because the majority of stroke patients are older and susceptible to sores developing. It wasn't pleasant at all but I had become a bit more hardened to people seeing my body and touching me where I don't necessarily want to be touched. I gritted my teeth and let them do the checks knowing that I had been extremely lucky to avoid any pressure sores to date. I saw Nula lot over the coming weeks and she was actually lovely. It just seemed like she was getting on with business when I arrived. Perhaps she felt it would be easier for me to just crack on. She was probably right.

It was late in the evening – after supper, when I arrived and I was really tired. It had been a long day getting ready and waiting for the ambulance. I just wanted them to leave me to sleep. I was also panicking because I knew Mum had dropped off the rest of my stuff but I didn't know where it was and I wasn't able to see her so knew she'd had to go straight home. It was a really stressful experience!

Once the nurses left I was able to sit and take stock. There was a poster on the wall saying that patients were to avoid 'pyjama paralysis' by making sure they got dressed every day. I had been washing every day obviously but I hadn't been distinguishing between night and day clothes. It felt quite scary to suddenly be on my own – I felt a bit like a child being dumped at secondary school with higher expectations of my ability.

It was so good to not be in a room with other people though. That made it all worthwhile. I was finally able to shut the door and keep the world out. Truth be told, it was quite weird that first night. Nula gave me my bedtime meds and I went to sleep. There wasn't really time for much else!

The next morning I woke up and it still all felt very strange. I had breakfast and was helped to get washed and dressed. I had a routine in hospital but I had to start again in rehab. I told them that I was hoisted onto a commode and then wheeled into the bathroom. They were fine with that and did the same. It was such a relief that I wasn't going to have to go back to being washed and dressed in my bed.

I had my first session of physio and I was really impressed. Alison was obviously extremely experienced and knew exactly what she was doing. I liked her because she gave me confidence that she would be able to help me. They had their own standing machine and we used that to do a little bit of standing but I told her I desperately wanted to be able to transfer from a wheelchair to a toilet. That's all I wanted and my expectations didn't go any further than that! She suggested I try a slide board where the edge is slipped under you and you then slide yourself over. It was tricky but I gave it everything and managed to do it.

Alison gave me a slide board to take back to my room. I was absolutely thrilled because it meant I didn't have to be hoisted from the bed or the chair. She got me a wheelchair and I was able to transfer (with support) from the wheelchair to the bed, to the toilet etc. It was such an enormous relief! I was able to wheel myself around my room a bit too although my arms were really weak and there wasn't that much room. I honestly can't even describe the relief I felt knowing that I could transfer to the toilet and from the chair to the bed. It was everything I had wanted for what seemed to be so long and that first time I wheeled myself into the bathroom and transferred was like all my Christmasses had come at once.

The next morning I woke up and a nurse I hadn't met called Sandra bustled in to get me out of bed and onto the commode to be wheeled in the bathroom. She bought in a standing frame – a bit like a trolley used for transporting large boxes and packages. I told her I couldn't stand up very well yet but that I had just the day before learnt to transfer using the slide board so would like to do that. She wouldn't let me use it and insisted that I should be using the standing trolley. I kept telling her I would need a lot of help to get to standing and I wasn't quite strong enough – I could use the slide board almost independently.

"But this is rehab, we have to progress you on," she said.

I couldn't believe that after being so happy and relieved about being able to transfer without help just the afternoon before, that was now not allowed just a few hours later. I got really upset to the point that I was absolutely sobbing. It was a stand off and I didn't have the energy for it. I don't know if they didn't really know what I had been through but my emotions were shot. I'd well and truly had enough. I had no one familiar to offer any comfort, was a little disorientated and very unsettled. Having just discovered the freedom of transferring myself it was cruelly snatched away and I had no control whatsoever.

In the end Sandra threw up her arms and said fine use the board but by then I was so upset and frustrated that I told her to just use the trolley but they'd have to help me to stand and then wheel me on the commode. That's what we did and standing was really tough. It took three people. But I knew I could do it by myself with the board so it was really disappointing.

I think back to this incident and I can still feel the devastation. It's such a simple thing but to me at that time it was absolutely everything. Why did she feel it necessary to enforce a standing frame and commode onto me? When it made me so visibly upset, why was it allowed to become such an issue? It was my choice. Mine. I should have been allowed to be in charge of my own recovery – it isn't a 'one size fits all' situation and I wasn't a 'normal stroke patient'. Not that there even is a 'typical stroke patient'. No one had the right to lead my recovery except for me and my wishes were made to feel unimportant. But I was alone and vulnerable in an unfamiliar place with unfamiliar people and didn't have the strength or the mental capacity to complain formally.

Sandra came back a bit later and told me that Alison is a friend of hers and she called her on her day off to ask whether I should be using the standing frame or the slide board and Alison had just said I can use either. I felt really guilty that she was disturbed on her day off. I couldn't see what they didn't understand about me wanting to hold onto the first bit of independence I had seen in months. When I had finished washing a different nurse let me slide back onto the wheelchair.

The next day when I had physio I spoke to Alison about it and got upset again. She said they were just trying to help me to get stronger. I didn't have the words or the energy to explain because in theory of course I wanted to stand and walk but in practice I had been totally dependent on other people for so long that having that tiny bit of independence taken away had a massive emotional impact. She said I could use the slide board for now.

I still hadn't mastered turning onto my side in bed so I asked Alison to help me and when she took me back to my room we tried it. Turned out that just bending my legs helped massively with getting them onto the bed – why hadn't I figured that out?! She then helped me to learn how to use the little strength in my legs to help turn my whole body so that I didn't need to be turned in bed anymore. Each tiny bit of independence was massive for me at this point and I was really pleased to be able to do this.

We worked out quite quickly that I was breathless on exertion with or without the oxygen and I tried my hardest to avoid using it. In the end it wasn't actually helping me to recover from exertion any quicker so we ditched it. Movement without having an oxygen tank trailing behind was much easier. Gradually we used the power assisted standing frame to build my leg muscles before I finally became able to take a few steps with a walking frame and Alison pushing on my hip and holding my left leg straight. Just a few steps to start with and the aim was to walk across the gym which I eventually managed after a few days. It was a massive accomplishment. I was so happy that I had been able to walk a few steps – but I was exhausted!

I was using a bike every day too. Sitting in the wheelchair I could sit in front of the pedals and challenge myself to get further and faster. Surprisingly I quite enjoyed it. It wasn't fast or intense but it was something I could challenge myself on.

I was starting to become more independent now as I was able to get myself from the bed to the wheelchair to the toilet and back and I started using the walking frame to transfer. It was huge for me! I got a bit overconfident walking with the frame and had my first fall one morning after washing. I'm still not too sure what happened but I think I twisted my body to change direction and my weak left knee just collapsed. It happened in slow motion as these things usually do and I was really shocked.

As soon as I hit the ground (I had a carer in the room with me) all the alarms went off and a hoard of staff came rushing into my room where they got me up. I didn't hurt myself thankfully but I knew I needed to be more careful.

21

AMBIDEXTROUSLY

A few days later Alison walked past my room and I was trying to get past my bed to my cupboard in the wheelchair. I could not get it past. There simply wasn't space. All I wanted was to see what clothes I had in there. She came in and rearranged my whole room so that the bed was in the corner rather than the middle of the room and I was able to move around and reach everything with the wheelchair. I chuckled while she was doing it – it felt a bit naughty but I was so pleased. Finally an improvement that was tangible. I could reach everything using the wheelchair or the walking frame. The bed was now behind a wall so was actually more private as I could sit in it and not see anyone at all if I wanted to.

I was slowly becoming more confident and was able to walk down the corridor for a short way using the walking frame with Alison walking next to me, another physio behind me with the wheelchair and one in front filming! I was so proud of that walk. It was the first thing I was able to send to my family that showed I was getting somewhere! Even though I only made it a very short way down the corridor – a matter of steps. I began to challenge myself to get a little further each time. It was only a few steps but I did manage the whole corridor by the time I went home a couple of months later.

After a few days I was getting more practiced about walking

short distances and suddenly I was passed over to a younger physio called Polly. I didn't see much of Alison after that. It seemed she had got me on my feet then moved over to a new person. Fair enough – I'm sure they had different cases needing the senior treatment coming in all the time and she was certainly very competent. I would have probably benefitted from being introduced to Polly rather than her just showing up one day with no warning but they didn't know about my apparent aversion to change just yet.

After that Polly and I worked on stamina by doing craft activities whilst standing at a bench for short times as well as side stepping around the bed. The crafting was actually a really clever move because it distracted me and I probably stood for longer whilst I was doing it. Lynsey had bought me a stone picture kit and I was making those whilst standing for increasing periods of time. It was exhausting but I enjoyed a little bit of crafting. It was refreshing to be able to bring a little bit of my life from before into helping me to get stronger. Crafting was always something that I very much enjoyed, even if I was now restricted to using my left hand.

I developed a little more strength and we tried to improve my breathing. That was one of the worst parts. Being unable to get enough oxygen is terrifying – I still get it on exertion now (my version of exertion which is moving from the kitchen the living room). It's like having someone's hand over your mouth whilst trying to take a deep breath. Only none of my breaths were actually deep. The damage to my lungs has improved but will most likely not recover fully. I'm more used to being breathless now but when you first start moving and you are unable to breathe properly it's really quite scary.

My day in rehab consisted of waking up early – I was woken at 6am, sometimes earlier. Medication, breakfast, get dressed and then wait around for my half hour physio session – which could have been at any time. There didn't seem to be a particular timetable – at least not for me. They most likely have had their own timetable but I like structure and would probably have benefitted from knowing. After physio I was back in my room on my own again and that was my day's activity done I watched a lot of tv. I did some colouring books and decided that as I still couldn't use my right hand very much, I would continue to learn to write with my left.

So I did lines as if I was at school with my left hand. I actually got fairly good at it and still only use my left hand to write now. I was able to move my right but I was, and am still, unable to feel my index and middle fingers. My thumb has little sensation and all fingers can also be really painful. It affects the strength in my arm as well as my ability to grip with my hand. The nerve conduction study previously had said that I had carpal tunnel which was contributing to the pain and weakness. I knew I could get that fixed but they said it wouldn't be anytime soon after being so unwell.

Carpal tunnel wasn't something I had before so I didn't know which pain and numbness was caused by that. The nerve conduction studies had also said about the critical care polyneuropathy so it was assumed that this was the reason for the arm problems. It's strange because you always assume that if you can't feel a limb, you can't move it. But I can move my right arm. It took a lot of work but, other than lifting it right above my head which I can't do, the movement is there. But it's weak. I can feel a channel on the outside of my arm all the way up and I can feel the back of my hand. The pattern of numbness is quite specific. My whole arm can be very painful.

I kept asking Polly to look at my arm and she said she would but she never actually did. I was a bit disappointed that no one seemed to be interested in what I wanted and I quickly gave up on my arm. They wanted to get me walking and that's what we did. My arm was ignored apart from one time a few weeks later when one of the OT's massaged it with some moisturizer – she did that once in my two to three month stay. That was the only time that anyone spent any time looking at my arm even though I kept asking.

But walking did improve. My stamina picked up until I was able to walk (unsteadily) small distances up the corridor with the walking frame and then (because I hated the frame) with crutches. The breathlessness was still a significant problem and meant that I had to keep stopping to catch my breath. Sometimes I was so breathless that it was quite frightening. The physio just kept telling me to breathe through my nose which was impossible because I felt like I wasn't getting enough oxygen. My room was as far from the gym as it could have been so getting from the room to there was an aim and it did take a long time for me to master it. By the time I left I could just about walk down the corridor – breathless but determined.

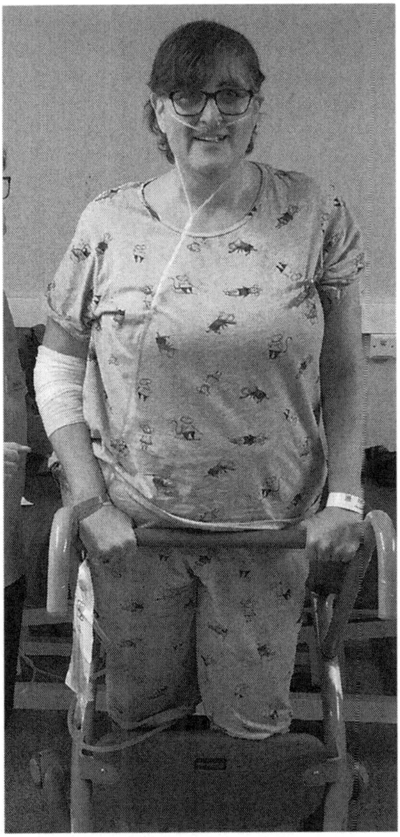

Standing on my own

22

FINDING FRIENDS

Dr. Lark from the stroke ward was also the consultant for the rehab ward so visited once a week. She had gone on holiday for two weeks after telling me I would be a full time wheelchair user and when she returned I was up on my feet with a walking frame – short distances but on my feet all the same. She actually had tears in her eyes because she was so pleased and surprised. It was lovely that she seemed to care.

Polly was suddenly moved to a different ward and I was allocated a new physio called Adam. Once I was able to do a few steps with supervision, Adam suddenly decided to take the wheelchair away. I was surprised and couldn't really understand why he was doing it but he told me it would be just in the corridor if I wanted it.

Here's the thing. I had become able to transfer from my bed or chair to the wheelchair and take myself into the toilet where I could then transfer and I was able to do all of this by myself. I couldn't, however, walk out to the corridor to get the wheelchair that was apparently being left there for me to use because it was too far on my own. So once that was taken out of my room, my tiny bit of emerging independence had just gone. It was extremely isolating. I was able to walk with crutches to the toilet but going anywhere else was totally out of the question so I was stuck in my bedroom. I know I had achieved all that I wanted very quickly and I realise that it's part of rehab to be moved onto the next step so encouraging me to use the wheelchair more could have stopped me from walking. I was so grateful for what I had achieved with their help but I was also gutted that the wheelchair I had been learning how to use was gone. It gave me a tiny bit of freedom to leave the room and even the ward – even though my arms were still very weak and it was difficult with my right arm being numb and unhelpful.

Soon afterwards, they gave the wheelchair to another patient and it remained in his room for the entire duration of his stay – even when he could walk right down the corridor with a stick.

I still don't understand why using a wheelchair wasn't part of my rehab. I was never going to leave on my feet and be back to normal so I find it strange that I wasn't given a chair to get used to wheelchair life – or at the very least referred to wheelchair services to be measured/assessed for one. I was just told there was a twelve-month waiting list and I would be turned away anyway because I could walk a little. I was eventually referred a few months after leaving rehab and waited just three weeks to be measured and another eight weeks for the chair to be delivered. A couple of years later I was moved into a powerchair.

Anyway, whilst in rehab I bought another wheelchair and Mum brought it into the hospital for me. I don't think the physio team were impressed but I was desperate to start being independent and, to be frank, I was past caring about what that looked like! I needed to get out of that small room, and eventually the ward, somehow. I didn't know what I was buying and bought an off-road wheelchair with thick tyres. Well it looked like the coolest option so why wouldn't I?! When I tried it, it worked fine but it was wider and it was a little awkward to get through the doors. I was a bit disappointed because it looked fun but would prove to be unsuitable.

Don't get me wrong, I am hugely grateful that I'm not a 100% wheelchair user but my mobility is still very limited in terms of breathing as well as foot drop from my numb left foot/ankle and my weak left knee. Walking with a numb foot is extremely difficult and requires absolute concentration which is exhausting. When you can't feel it, it feels as if you have a massive ball of cotton wool instead of a leg and you have to find ways to move that effectively. I figured out that if I kick out my leg after lifting it off the ground, I could see where it was going to land and prepare myself but my leg was heavy and difficult to swing. As long as I could see it, wasn't too fatigued to lift it properly and was concentrating, I could more or less control where it landed on the ground.

The breathlessness remained – and remains, a significant problem. Unsurprisingly, my lungs are scarred from the Covid and the sticky secretions that accumulated inside. I hope to be able to hone my breathing skills – a consultant later on told me I need to learn to breathe all over again in the same way I have relearnt how to walk. There is room for improvement still but my lungs are smaller now and my diaphragm has lifted on one side so there are structural changes as well as the scarring. The breathing difficulties add to my mobility constraints significantly.

One day I heard someone new arrive and she was put into the room next door to me – sharing my bathroom. I heard her say she had been paralyzed from the waist down. She sounded young and I hoped I might get to meet her at some point. A few days later I attempted to walk out of my room by myself because I was fed up of sitting in there on my own. I didn't get far – just slightly past the room next door but I noticed the lady who had arrived was in there on her own so I popped my head in to say hi. We were sharing a bathroom so it seemed only polite!

Her name was Janie and she was a little older than me but

not much. She had unexpectedly been paralyzed and was not able to walk at all. She was learning to live all over again. She was insanely positive and fun to chat to. It was nice to speak to someone for a change. Her partner was also there and thank goodness he was because I sat on a chair to talk to them and got stuck on it. It hadn't occurred to me that I couldn't get up from a chair if it had no arms! He had to pull me up and he then helped me back to my room.

Janie and I chatted quite a lot whilst we were in rehab together. We both had our different challenges and frustrations but we tried to look out for each other. We developed a bathroom sharing system and often called to each other when we were in the bathroom. It was like a comedy sketch but the system actually became quite effective and we shared successfully. We still joke about it now.

One morning an OT (occupational therapist) I had never met called Sharon came in and informed me that I was doing a 'wash and dress' that day. They put a sign on my door to indicate that it was happening. I didn't even really know what that meant but the nurses said I couldn't get dressed because of it. I didn't like the sound of it actually. A bit later Sharon and a different OT that I had never met bustled in and told me they were going to observe me washing and dressing myself. I was a bit taken by surprise because no one had explained this to me. No reasons given or reassurance, just matter of fact.

I was watched getting myself to the bathroom and then had to strip wash. Sharon stayed with me and the other was getting my clothes etc ready outside the bathroom. Now this is clearly something they do to assess how independent a patient is but, as far as I was concerned, they were just watching me strip wash. Had I been myself now I would have asked them what the aim of the exercise was and then probably just got on with it but I was extremely vulnerable and not myself at all. So I just found it really distressing. I was pretty fed up of people wanting to do things to me or watching me do things that made me uncomfortable and I was now conscious enough to voice my disapproval!

I know I was probably a little obstructive but I was massively uncomfortable and at no point did anyone ask me if I was okay with them being there.

"I just want to go home," I said and Sharon just said, "We're trying to help you get home!"

"No, I mean I want to go now."

To which she said, "that's fine. You go home with a catheter and needing to use a commode to wash yourself then."

It felt aggressive and confrontational and, already uncomfortable, I continued to sob my way through until I was finished washing. There was very little empathy or comfort available and it eventually ended.

I can actually still feel the vulnerability of having to wash intimately with someone observing but not considering my feelings about it at all and not responding to my distress. I had no warning and no explanation. I was still struggling with what had happened to me and with the weaknesses I had been left with. I needed someone to understand and show some empathy.

I think a lot about stroke patients who are in rehab having to go through this. A stroke is destructive enough without this on top. I accept it needs to be done to assess a person's abilities, however I think it's so important to consider the individual not the 'stroke patient'. I was fully aware of everything. There were no problems with my cognition and understanding. My communication was effective. But, despite the obvious distress that it was causing me, the procedure continued and no comfort was offered. Stroke patients are vulnerable and this type of treatment literally strips away that person's identity, putting them firmly into a box. Of course in my case, my stroke was probably the least of my problems, but all that aside, I was obstructive therefore I was treated like a 'troublemaker'. A naughty child.

Unfortunately, because of Covid, visitors still weren't allowed and I was very isolated. All the communal areas of the ward were closed (not that I'm especially 'communal' at the best of times) so there were no opportunities to do anything much at all. As a result I spent 98% of my time alone in my room. I'd been saying I just wanted to be alone when I was on the stroke ward and I truly meant it. I didn't complain about not seeing anyone else much but thinking back, I probably could have used a bit more help.

Eventually visitors started to be allowed but only one per patient at a time. They had to book a slot to come in to prevent too much of a crossover of different people. So obviously Mum came but also some other friends who came on different days.

Lynsey had been planning to come over on her way home from seeing a show in London and she was going to bring me some more gifts and leave them at reception. Just two days before she was due to come they opened up the visiting to the one person per patient. I called her to tell her that she could come in and she cried. It was so amazing to feel loved.

When she visited, poor Derek, her husband, had to stay outside in the car. She stayed with me for quite a long time and we even went outside with me in the wheelchair. The sky was blue and the sun was shining. Other than the ambulance journey from hospital to rehab I hadn't been outside for weeks. Breathing fresh air was so refreshing – not that I've ever been a fan of sunshine but it was so nice. We obeyed the rules and Derek stayed in the car but I was able to wave at him. It meant such a lot knowing how far away they live but they made the effort to come – in fact they didn't give it a second thought.

For a long time, even when I was walking with crutches, I still wasn't allowed to go to the toilet by myself. I had to call a nurse in to walk with me. I was desperate to just be able to take myself to the toilet and eventually the physio said I could. That night I came out of the bathroom and had just put my pyjamas on. I walked over to my cupboard with my clothes in my hand, bent down to pop them in my washing bag and before I knew it my body was twisting to the side. I tried to save myself but I went down with a very loud clatter. Everyone came running and the alarm sounded again.

I landed, hitting my head on the metal frame of the bed. It hurt so much! Unfortunately I had hit one of my shunts very hard and it swelled up very quickly. A cold compress was put on and a lovely care assistant called Sam was straight on the floor with me. She was so caring, I was crying and she was just holding me and talking to me.

I needed to get up but I knew there was no way I could get up from the floor. It took a few of the staff, a stool for me to sit on halfway and a lot of help to get me from the floor to the chair. I felt really unwell where I had hit my head so hard. It was a Sunday so they had to call the on call doctor who came to see me after a couple of hours. She asked me if I wanted to go to A&E and of course I didn't! She told me to get some sleep so I did.

I woke up the next morning and, being Monday, the ward doctor was back. When she found out about the fall she came in and told me I needed to go and have a scan. I couldn't believe it having stayed there overnight after the head injury. She was insistent that because I was on blood thinners I had to go. So I reluctantly agreed and she said she would arrange transport. I asked if mum could take me but they said no.

I waited for the transport. And waited. I missed physio

– partly because I was feeling too poorly and partly because I was expecting transport and didn't know when they might turn up.

I waited a bit longer, slept and waited more. It got to bedtime again and I still hadn't been collected. The nurses chased transport but just got told they were busy. So more than 24 hours after the head injury, I went to bed for the second night. One of the night staff came in and I mentioned I was waiting for transport to take me for a scan. She said that once I got there I would need to wait my turn anyway. I didn't really understand as I was told I was going for a scan. In my head, transport would collect me, take me to the scanner and then bring me back. But she said I would need to wait my turn in A&E and they were very busy.

I didn't expect special treatment. I had no idea that A&E was where I was going! When I realised I was going to be taken into A&E and left there on my own I got really upset. I had just almost died from infection, was on immunosuppressants and steroids and was extremely vulnerable to further infection. I was also incredibly anxious and unsettled. Another nurse reassured me that it was wrong and I was just going for a scan, not A&E. She told me to go to bed and I was exhausted so I did.

At 3am I was woken up. Transport had arrived. 3am. I was got out of bed and put into an ambulance in a wheelchair. I was obviously still a bit lost and confused with everything that had happened and I had a head injury so this was really scary for me. I asked if I was going to have to wait that many hours to get transport back but no one could tell me. I told the ward in no uncertain terms that if I had to wait as long to get transport back I would be calling Mum to come and get me and then take me home. I couldn't bear the thought of being sat in A&E for that long waiting to come back! They told me that I was under their care so they hoped I wouldn't.

As I was getting into the ambulance I asked them if I was

going to A&E or just for a scan and they said I was going to A&E. I told them I was really scared and nervous of how vulnerable I was – given everything I had been through. They were very sweet but couldn't really reassure me.

When we arrived in A&E I was taken in and the paramedic told me that they were full and I was going to have to wait in the corridor. She said she would find me a trolley to lay down on. I got onto the trolley and was left in the corridor. I can remember looking around at everyone questioning in my head whether they were infectious and I did get upset again. I really wasn't well with concussion and shouldn't have been there on my own. In hindsight, should I have expected a staff member from rehab to come with me? I accept they were busy but I was under their care and I could have just gone anywhere – I seriously contemplated just going home! I was an inpatient – surely they had a duty of care to keep me safe.

The staff in A&E were stretched but they were really lovely and when they realised I was upset, unwell and vulnerable to infection they found me a cubicle to wait in. I felt a bit safer in there and didn't mind waiting. They were exceptionally busy!

I had a scan and then just slept a lot in my cubicle while I waited for the doctor. I didn't have a pillow so had to roll up my jacket to put my head on. I still had a catheter in with the flip flow valve so needed to empty it from time to time. I was just starting to get mild sensations of needing to empty my bladder by this point and I knew I needed to empty it.

When a care assistant came in I asked her for a bottle to empty my catheter. She said no. I should get up and go to the toilet. She then tried to get me out of bed to walk to the toilet. I told her I couldn't walk without crutches and even then I could literally only take a few steps but for some reason she wouldn't believe me. I told her I had been in hospital for months and she told me that was even more reason for me to get up and mobilise. "You need to keep your legs moving! They wouldn't have discharged you if you couldn't mobilise!" she said. I was getting a bit frustrated by now and explained that I hadn't been discharged and was still an inpatient until eventually she listened. She tutted and left to get a bottle – which someone else brought in a while later.

Why are there a handful of care staff who either don't listen or don't seem to care? I didn't encounter many in the long time I was an inpatient – the majority of the staff were amazing but there was the odd one or two who, for some reason, failed me. I find it difficult to understand why these people choose to work in this type of setting where they need to genuinely care. Where people really need them. I do also understand that some patients need the 'tough love' approach but I wasn't one of them so it was misguided and counterproductive.

Anyway, the doctor eventually came in at 7pm and said my scan was clear and I could leave. They arranged transport back which was surprisingly quick and I went back to rehab. I'd been gone all day and the final insult was that I hadn't been there to reserve dinner so I had to have what was left. I didn't eat much that evening – I was exhausted anyway.

It took me a couple of days to recover from the head injury and disturbed night but when I was feeling better I started some physio again. Adam and I began to work on balance as that was clearly the reason for my fall. I also asked to practise getting up off the floor – I knew that if I went home and fell Mum wouldn't be able to get me up again so I needed to be able to do it.

The balance exercises really helped and I did get more steady but try as I might, I just couldn't get up from the floor! I had to kneel on the floor in front of the bed, lift up one leg and then push up on the table. Easy right? Well my legs were just useless and I couldn't get the muscles to tense – even after all that time! In fact I fell sideways whilst trying and ended up sitting on the lap of the young, fit, male physio helping me! I think even he was a bit surprised.

So we tried kneeling on the floor then putting my feet down as if they were on tiptoes and pushing up with both legs together. Well the trouble was, I couldn't feel my left foot so getting it into that position was impossible and pushing up on it was even more so! They demonstrated it a few times and it honestly looked so easy but I could not do it. Being able to get up from the floor was a condition we agreed on for me being discharged. Three years later and I still really struggle with it. Lucky we didn't stick to that agreement!

23

RULE BREAKING

When it came to Andrew and Phil, aside from the brief hello I had with Andrew whilst I was on the stroke ward, I hadn't seen them for months. I could see lots of visitors coming in and people often seemed to have more than one visitor. So even though the boys planned to come inside one at a time, they were both there at the same time when they visited. In the days leading up to their visit I did ask a few staff members if they could possibly make an exception to the one person only rule. I had been in hospital for months and we are such a small family that seeing them was so important to my mental health. Some said they would check but never came back to me and others said just do it, no one will notice. No one said no (but no one said yes).

We broke the rules. I was just so desperate to see them after everything and I thought it would be okay – we didn't hug or go near each other. Thing is, they live in the same house and they both work from home so I knew the risk of Covid was fairly low – and equal between them regardless of whether they came in together or separately. It made perfect sense in my head.

Andrew popped out to get a coffee for all of us and when he came back the ward sister Dolly came in and she was really cross that I'd had two people in my room. She said that if Covid came onto the ward then we would know which room it came from and if I got Covid I'd have to be moved to a Covid ward and would need to leave rehab. We all felt totally awful – I knew it was against the rules but I was so desperate to see them that maybe it clouded my judgement. When they first arrived we even asked to go outside thinking that being in the fresh air would mitigate the risk but Dolly had said I wasn't allowed out of the ward – despite Lynsey taking me outside just the day before – with permission. It was all a bit confused. There were no communal areas open due to Covid so it was my room or one waits in the car.

When the boys left I sobbed into my pillow for ages. That greatly anticipated and much looked forward to visit that meant so very much to me was tainted – we aren't natural rule breakers.

Anyway, the very next day I found out that the nurse who had been looking after me that day had Covid. In fact quite a few of the staff had it. Turns out it was already there and it certainly didn't come from my room. They struggled with staffing for quite a while after that as a lot of staff started to catch it. I don't know if any patients caught it but luckily I didn't. They were very good about wearing masks and obviously did a good job with infection control because I didn't even catch it from the nurse caring for me!

There started to be a lot of agency staff because of staff sickness and staffing levels appeared to be pretty low. I saw less and less people during the day because the staff were spread thinly across the ward. Some staff liked to chat and would come in talking to the point where I was exhausted but generally they just didn't have time. It was such a lonely place. I loved it when they came in and had a few minutes to talk to me. Some of them were wonderful and I enjoyed talking to them.

On one sunny afternoon a while later I was sitting on my own in my room as usual and Janie came past the door. She told me to get in the wheelchair and go with her. We had been outside to the garden together a couple of times – I was much weaker than her and couldn't push myself as well as she did so she helped me by pushing me with her feet whilst pushing herself with her arms! Anyway on that day I went with her and she told me that she and her partner were going to the pub and I was going with them! I was surprised but went with it.

Unfortunately I was too weak to push myself down there so Janie's partner had to push me and she rolled herself all the way – super-Janie! Her skills were seriously impressive after such a short time in a chair. Apart from when she was going downhill, lost control, bumped down the kerb and flew into the road – where luckily she stopped behind a parked car but I was certain she was going to fall out of her chair!

When we arrived at the pub I suddenly felt an instant regret. I was worried about how I was going to cope – with being around other people and the risk of infection as well as with the additional stimuli I was suddenly facing. But I had a drink. I drank it quite quickly because I just really wanted to leave. I didn't expect to struggle quite so much and I felt really guilty because Janie needed to get out and was enjoying herself. For a while I sat wondering if I could push myself back so they didn't have to leave with the party pooper. But I knew for sure that I would never make it and then I would be stranded by myself. So we all had to go back because I just couldn't cope. Janie was so understanding and sweet but I know she would have liked to stay. Apparently she got told off later for leading me astray. We had told someone when we left but it seems the message didn't get through and they didn't know where we were? No one ever said anything to me so poor Janie fully got the blame!

Janie left just before me and I missed her being there. It felt strange not having anyone to call out to before going into the bathroom. It was only a couple of nights after she left that I went home though so I didn't have to miss her for too long. Janie had to go onto a spinal unit to rehabilitate further and returned home a couple of months later.

There were still patients who shouted all night. Most were still elderly and many were understandably confused. The difference was that I was able to close the door and block it out to an extent. There was one lady who would just wheel herself up and down the corridor in a wheelchair. She was elderly and I think she was a bit confused but she would roll herself back and forth for hours. Sometimes she just sat in the wheelchair in her doorway. I'd see her when I went past for physio. But suddenly she started rolling into my room and sitting in my doorway. Facing outwards. I'll be honest, I felt sorry for her but I also felt a bit uncomfortable for some reason. It felt like a bit of a privacy invasion. I'm pretty sure she thought she was in her room but was just lost.

After a few times and as my bed was behind a wall I decided to just get into my bed and ignore it. Pretend she wasn't sat there at all! I just went off into my own world as I did so often. Staff moved her on when they noticed her.

The ward doctor Dr Johnson was quite to the point and said things as they were. She would come each day and then once a week would do a ward round with the consultant. I'll admit I didn't have full confidence in her after the junior doctors I had encountered but one day she came in with Dr Lark and we were talking about what I had been through. They left and obviously continued to talk about it some more as a short while later she came in with tears in her eyes to tell me I was doing an amazing job and that I should be proud of myself. She did really care and after that I enjoyed her daily visit.

Unfortunately the junior doctors only stay on one ward for a certain amount of time and she was replaced with another who did not fill me with confidence at all! She would walk around with her handbag strapped to her – in fact I would go as far as to say that she strutted around! She seemed less knowledgeable than I am. I didn't see her that much but when I did it was brief and she wasn't really interested.

After a month or so in rehab, it was decided that my catheter could be removed. I was really scared about having it removed because I wasn't completely sure I was feeling enough of a sensation and I didn't know how it was removed. It actually pulled out remarkably easily just down my trouser leg and I was then just left to it. I definitely overthought the whole process! To start with I went to the loo regularly just to make sure but once it was time to go to bed it felt slightly more scary! I was terrified of wetting myself during the night if I didn't feel any urge to go. In the end it was okay. I woke up to use the loo once and then when I got up in the morning I didn't quite make it but gradually I had to retrain the muscles to do what I had been doing since I was about 3 years old.

Unfortunately myself and Janie next door both had our catheters removed at the same time. I had mine for longer but I also had longer with the flip flow valve so I did have some control. Poor Janie didn't have the flip flow for long and wasn't ready to have hers removed. She really struggled. But we shared a bathroom so I guess you can imagine the situation. When she was in the bathroom she locked the door on my side and I did the same when I was in there. Either one of us could easily forget to unlock when we had finished – Janie was reliant on the carers doing it and I was simply reliant on my useless memory. So a few times we would go to use the bathroom and be unable to get in! Time is of the essence when you've just had a catheter removed so, well you can imagine!

There was a psychologist in rehab called Sally. Her specialism was stroke patients so she wasn't necessarily equipped to handle multi systemic issues and PTSD but I enjoyed talking to her and felt quite comfortable around her. She learnt fairly quickly that I am very self critical and she referred to the 'inner critic' regularly. It is like a nagging voice in the back of your head which makes you doubt your own thoughts and feelings, replacing them with hyper critical interpretations which start to eat away at the way you think about everything. Is everyone laughing at me? Do they think I'm putting it on? Not recovering quickly enough? Not trying hard enough? It's draining when you are trying to contend with rehab.

I saw her fairly regularly and talking to her helped. The inner critic has always been there and will always be there. I am more aware of it and subsequently more able to challenge it but it is still a strong influence. I found it very difficult because in my head the physio team were laughing at me behind my back and thinking that I wasn't trying. Thinking that maybe I just wanted to be in a wheelchair, maybe I wasn't bothered, was lazy, wasn't interested in getting better. I'm sure they weren't thinking that – I hope they weren't but that voice in my head is strong. Same with the nurses. They were trying to avoid being my nurse. Didn't want to be around someone who cried so much or thought I was just a complete misery who didn't want to get better. I'm fairly sure they didn't think that but my brain has this superpower which can blow any comment, action, situation etc completely out of proportion and make it explode into a massive issue in my head. Knowing that sadly doesn't prevent it from happening – although it does make me more able to understand where the thoughts are coming from and also gives me tools to minimize the destruction that the inner critic can (and does) cause.

She encouraged me to write a letter to myself because I was showing myself a lot less empathy and understanding than

I would someone else. It helped to think about how kind and understanding I would be with someone else but I didn't apply that to myself.

Dear Jen,

Wow! What a journey you're on!

When I first heard you had Covid, I wasn't overly surprised. We'd been dealing with it for many months and it was always there. I knew you were vulnerable given your lupus and your medications but I didn't really know anyone who had been really sick with it. We all think it will never happen to us don't we. It's so easy to be complacent about it.

When you let me know how unwell you were though, I became a little more concerned. I know from your messages that you were really struggling to breathe and bleeding internally. It sounded scary.

I know you hate to make a fuss and think that everyone is thinking it's all in your head because you've been told that all your life but this was real. Being sent away from A&E when you first went must have made you believe this even more. We will never know if things would be different had you been taken seriously this time. It is unimportant now.

What happened next was scary and you were on the brink of death. You don't remember the worst parts which is a blessing but it did happen and it did happen to you.

I know things have been very difficult since you woke up and that many things have presented challenges – personal care for example, and getting moving. I know there has been physical, mental and emotional pain at every step of the way. You gave it everything and got on with what you needed to do despite this.

The future was, and remains, uncertain and you have accepted this bravely, whilst working hard to shape the best future you could achieve. It most certainly isn't the future you imagined for yourself but you have tried so hard to accept that and stay positive. You were told that you would be a full time wheelchair user and you took this really calmly – but stayed determined to walk. And you did just that! You never asked 'why me'?

Just a few months after you woke up you were on your feet (if a little wobbly) and able to walk – even if you can only really walk inside the house. You have still far exceeded everyone else's expectations even if you didn't meet your own.

I know you still think that people are rolling their eyes and judging you when you use the wheelchair, thinking you are just being lazy but if they don't know that you would give anything to be back where you were before, then they don't know you at all and are, therefore, not important in your future.

Just keep remembering those first phone calls to your friends and family. Do you remember how they reacted? Remember when you called Andrew from ITU? He was so shocked he could barely speak! And the friends who cried when they first heard your voice. This is testament to how they would have felt had you not recovered. They cared.

Now you are moving into the next phase of recovery and I know it is causing you some anxiety but keep those people who cared in your head and lean on them because, even if you do sometimes wish things had been different and you hadn't woken up, they don't.

You have accepted the change in your mobility without complaining and you remain positive that you can develop skills to rebuild your life despite the obstacles. You know there will be difficulties but you also know you are strong enough to overcome them.

Just remember, people love you and would have been devastated if you hadn't made it. Be proud of how strong you are.

Love always, Jen

24

REDISCOVERING THE OUTSIDE WORLD

Sally the psychologist asked whether I would mind if her student did some cognitive testing with me. She felt it would be beneficial to see if I had lost any ability after the stroke. I agreed and the student started to visit. She did the cognitive testing across three consecutive days and it took around six hours. Although the distraction was great, I was totally exhausted every day! We did a variety of tests like memory tests, repeating words and phrases, finding my way out of a maze (on paper) etc. it was interesting but it was quite draining.

Once she had finished she typed up a detailed report regarding my ability. It was like an end of term school report! She concluded that my processing speed was slower and this was affecting my ability to remember things. She did get it spot on actually. It still does take me longer to process information and I quite often switch off. I have trouble watching television as it requires me to concentrate and I lose interest very quickly. I spend a lot more time just in my own mind staring into space. So, even though it wiped me out, it was a useful exercise. Hopefully it was useful for her studies too. The truth is, I am exceptionally lucky to still be coherent and able to think, communicate and complete tasks I was doing before. I could have lost that and being 'locked in' is just unimaginable.

As the days went by during the staff shortages, my mental health really began to suffer. Physio ground to a halt. The forty minutes or so per day of physio that I was having initially turned into twenty minutes every other day – if I was lucky. The rest of the time I was in my room on my own – apart from short visits from family and friends. I had the wheelchair that Mum had brought in but I still had very little strength in my right arm – well both my arms really and whilst I could roll down the corridor, I struggled to go anywhere further than that.

As I became lower and lower in mood I lost the motivation to do anything. I just wanted to go home and the staff shortage which led to the lack of physio unfortunately came at the same time so I pretty much gave up. I was able to walk to the bathroom which was all I had wanted and I genuinely just wanted to leave. The four walls seemed like a prison and the routine was ingrained in me. In between sleeping, washing and eating there was very little to get excited about. The physio I was having by then was very generic and I wasn't really gaining anything from it anymore. I had lost the will to fight and I think they had lost interest in me. The bar had been set low to start with and I had achieved more than I ever thought possible. I almost needed respite from rehab!

The last six months had truly been the hardest of my life and I was totally exhausted. Exhausted of fighting, of the pain, the tiredness, the boredom. Of being a 'miracle'. I didn't want to be a miracle – I wanted it to not have happened. Life was good before – despite the health difficulties I had, and I wanted that back. But I knew things would never be exactly back to before. Covid had thrown a wrecking ball through my life and I was still reeling.

I still didn't really know what life was going to look like

eventually. When I first arrived in rehab, Alison had told me I would be back to normal with nothing to show from it all. I would walk out of there. The trouble was I didn't believe that anymore and I didn't have any real motivation left in me. I knew I had let everyone down but I had spent a lot of time thinking about the future and I didn't have set expectations. To be as independent as possible was my only wish. I didn't really care what that looked like. I'm sure that sounds odd to you if you haven't battled serious illness. Accepting the wheelchair for now and trying to move forward instead of constantly fighting. The truth is I'm at peace with whatever life looks like and that's the crux of it all. Independence was all I wanted to achieve. Of course I sometimes feel sorry for myself despite this and it can be very frustrating but at some point you have to address what life is actually going to be like.

I was feeling a bit under the weather and couldn't quite put my finger on why but wondered if I might have a UTI. It wasn't long since my catheter had been removed so they took a sample for testing. It came back positive so I was put onto antibiotics and told to drink plenty. At the same time a blood test confirmed that I still had heart failure so I was then put onto restricted fluids! At which point I was suddenly thirsty all the time! The balance was difficult and even the ward doctor didn't seem to know whether I needed to drink or not!

One day, totally out of the blue, the staff suggested that Mum take me to a pub for a meal. It was made to seem like we should have been doing it all along but no one had told us! So Mum and I went to the pub and had lunch. It was hard with the wheelchair – we weren't given any guidance or help with getting it in or out of the car or with how to negotiate paths, car parks, any slanted surface. But we did it and it was lovely to eat something different and be with Mum away from the hospital. I coped relatively well with it even though it made me really anxious being away from the safety of the ward. I was utterly broken by the time we got back but it was good to just be able to go to bed.

As I was now allowed out, Mum brought the dogs over a couple of times and I went out to the car park to see them. I had missed them so much and I think the feeling was mutual because they were really excited – although they got bored of me quite quickly! It was so good to see them and I was so grateful that Mum had brought them to see me. I had totally forgotten that my eldest dog had been diagnosed with cancer the day after I tested positive for Covid. So not only was Mum worrying about me during the intensive care battle, she was also caring for Poppy who was having aggressive treatment. Luckily Poppy beat the cancer and went into remission.

Eventually, we started to talk about me going home. I spoke to Dr Lark because I was anxious about it and she said that we would maybe start slowly with a day or even a weekend visit to see if I was well enough. She called in Megan who was the OT looking after me at the time so she was able to be involved in the conversation as she would be central in organizing it all. We would wait until the time was right for me and do it slowly to make sure that everything was ready. They would get handrails fitted and my house would be assessed before I went to see what was needed. She promised that it would only be when I agreed that the time was right and that it would be done very carefully. I felt much better after we had talked about it.

We talked about the possibility of downstairs living to start with. I explained that we had one big room so downstairs living for me would involve sleeping in the lounge and washing in the kitchen because, whilst we did have a downstairs toilet, it was too small to wash in the sink there. It was decided that it probably wouldn't work and so in the physio sessions I did have, we started to look at stairs. I was managing them fairly well despite the breathlessness. Four was my maximum and I was breathless enough that I needed to stop. But I wanted to go home so that gave me a bit of motivation to go up those steps.

Then one day I was seen by a physio I had never seen before called Henry. He took me to the stairs and I went up and down four steps. He then took me back to my room and left me. A short while later he came back and excitedly announced that I was going home as soon as possible and would have to live downstairs. I tried to explain that I couldn't do that but he was insistent and then left. Yet again, I was upset. What happened to waiting until I agreed the time was right? What happened to the day or weekend release?

Sandra, the nurse I had clashed with right at the start of my rehab admission was there and asked me if I thought maybe I had OCD because I didn't do well with change. During my time in rehab we had actually put that first meeting behind us and she looked after me quite a lot. We understood each other after our rocky start and actually I was able to chat to her quite confidently.

But she was wrong on this. No, I do not like change generally (does anyone?) but I had almost lost my life, I had lost my mobility, my voice, my confidence, my job. I had lost every aspect of my life. I had got used to HDU after leaving ICU, then the ward (in various places) and then rehab. Each had been a massive change because my needs were so great and I wasn't allowed visitors so the staff had to become my family each time – more so because they were sometimes involved in the most intimate aspects of my care. I had to put all my trust in them because I simply had no choice. After exactly six months in hospital I was institutionalized and suddenly I was being lobbed out into the world with no idea how the hell I was going to live – with a wheelchair to adapt to, physical limitations that I hadn't needed to address so much in the confines of the hospital and a new life which I did not authorize nor want. No one had shown me how to use the wheelchair or helped to explain how I was meant to get it in and out of the car or generally manage life in it. I was unprepared for the outside world and I knew it. It was terrifying. I was also aware that Mum felt the same and that me being home would be really stressful for her.

So to then be told that I was going to have to work out a way to sleep downstairs, buy a bed and just manage was a shock – especially when I had been assured of this soft gentle discharge. There was no possibility of getting help with obtaining a bed for me to sleep in and I was told that the handrails would take a few weeks to go in.

Because I was upset Sally, the psychologist, came back to speak to me. She told me that staying put 'wasn't an option' so basically I had to suck it up and leave.

Mum started looking for a bed for me. We would have to lose our dining table and I would have to sleep in the living room. There didn't appear to be a choice. Other than just doing the stairs and taking the risk which I seriously considered. I told the psychologist that I would just do the stairs without a handrail because I couldn't sleep downstairs. She told me that they hoped I wouldn't take the risk. She said I was in danger of serious injury if I was to go up or down the stairs without a fixed handrail on both sides.

I also voiced my concerns that I would leave rehab and then be just on my own. I didn't have any follow up appointments on the horizon and it felt like rehab were washing their hands of me. I'm certain that wasn't the case but I knew that if I left without firm follow ups booked I would be left floundering. Who was going to coordinate my care? There was no one. I hadn't seen my

GP for a long time because of Covid so I didn't think she would be in a position to coordinate. I was reassured that there would be follow ups.

The rehab OT Megan (when she came back from having Covid) tried to organise handrails but when she realised I lived in a different borough she seemed to hit a dead end and told me she couldn't get hold of anyone but had left a message. I'm sure it didn't make me popular but I knew someone who worked in my borough doing adaptations and she contacted the OT herself for me and she was amazing. Solely because of her, a handrail was fitted in time for me to go home and we didn't have to cope with downstairs living.

And then after believing I was going to be sent home without any preparation at all, two OTs took me for a home visit where they checked I could do the stairs – it took a long time and was exhausting but I did do them. They also checked if I could sit safely in the chair that the local OT had arranged, that I could get up from the bed and the toilets. It wasn't the day or weekend visit that I had been promised but it was a good chance to reassure Mum that I was going to be okay. We went back to rehab and I was to be discharged the next day.

The evening before I went home I made my second cup of tea since becoming unwell! I asked to because I hadn't been given the opportunity to do that much apart from once or twice when I was watched by the OT. This time I just wanted a cup of tea and figured that if I was ready to go home they'd have to trust me with the kettle. I carried my plastic cup to the lounge (which was closed due to Covid) and made it. I couldn't carry it back to my room annoyingly as I was on two crutches but I still made it myself. It did make me wonder how I would be able to do that alone but I knew that Mum would be able to help me until I found my feet as it were. I packed what was left of my things ready to go the next morning. I had taken some of my things home already when I went for the visit so that I didn't have so much to bring home in the ambulance.

The next morning I was woken up as normal and got myself ready then waited for the ambulance. They didn't keep me waiting too long this time. As they wheeled me out everyone came to say goodbye. They took their masks off to give me a hug and it was so lovely to see their faces – even though they didn't look like I had imagined! Rehab had its faults and it definitely wasn't the right place for me – I needed a much more holistic approach as I had fairly systemic damage and the majority wasn't caused by my stroke – but I had built relationships with them. I felt safe there and, like everywhere I had been, they had become my temporary family. Going home was exciting but actually pretty terrifying.

The ambulance delivered me home and Mum was there to meet me. It was an odd feeling for both of us because she was obviously anxious to have me home. I was still very weak and had a lot of recovering still to do. We were essentially alone and it was a huge step. I should have felt happy to be home – and I was, especially to be with Mum. But at the same time, and I can't really explain it, I missed the institution factor. The routine, the people, the stimulation – even though in the later days it was very boring and lonely.

And then we were on our own.

STEPH

Jen finally came home on 29 April 2022. It was a shock and we began to realise just how far she had come, but my goodness what a very long way she still had to go – it was truly rather scary. It felt as if she had suddenly been dismissed, but Jen being Jen tried so very hard to be independent and worked tirelessly to get some normality back.

In hindsight neither of us had been truly prepared for this transition. We are eternally grateful to the people who worked with her, but Jen had not been happy in rehab. The focus was very specific. What about life skills or how to manage life with a wheelchair etc? There was also very little liaison with me during her stay. I was initially told that in rehab staff and family work together and in fact towards the end of her stay there was a meeting to discuss her homecoming – but it had already all been decided with no say for us. When I expressed my worry about how she would cope I was told, "Well you could have been taking her out to get used to using the wheelchair etc." I found this so upsetting – it made me feel really guilty that I hadn't taken her out. But in fact there was one time when she asked about going out and was told a strong NO.

Her time on the stoke ward was difficult for her. The physios and nurses were amazing but she was the only young person on the ward which was quite hard. Jen was also very disturbed by memories and delirium from her time on ICU.

This time on ICU was of course very difficult and frightening for all of us but I know that all the staff were wonderful and without any of them Jen would no longer be here. They were working under very difficult and unusual circumstances and gave their all. Yes it was sometimes difficult to speak with doctors and with no visiting allowed we were often quite unsure what was happening but I know they all did the very best they could and I am eternally grateful. Without their care Jen would not be here now.

Back in the outside world we both felt a bit lost. It was a very difficult time for Jen to work through! We were both unprepared for this life change I'm afraid to say. We hope that through telling our experiences we may be able to help others, indeed Jen has been asked three times already to speak with medical students and we hope that hearing her story can provide an insight through a patient's eyes.

Jen is a fighter who tries to put a brave face on for the outside world but this last challenge has taken its toll on her. It is taking time to come to terms with everything that's happened. As a family we never realised all the traumas that being in a coma etc and surviving against all odds would have on her and indeed the family. This is real life, nothing like it is in the movies where the person usually gets up and over it immediately.

I hope that Jen will eventually become more and more independent and begin to feel more comfortable as she is. Although Covid has definitely left its mark on her life – on all of our lives, there is no going back. Just moving on and making the very best of life as it is now.

Hopefully talking about our experiences will help others as time goes on.

25

FLOUNDERING AND FALLING

Surprisingly, I seem to remember more about rehab than about those first weeks at home. I can only think it was because everything was just so overwhelming at home. There was so much to think about and consider. Making decisions that were mostly made for me in hospital – even just what to eat, was tiring and generally having a massive amount of extra stimuli that I wasn't used to.

I think I was wholly unprepared for the outside world in every way. Being able to walk down a corridor is very different to being able to function outside of an institution. Other than those two cups of tea, there was no opportunity to do anything else for myself so it was a shock suddenly being faced with the sights, sounds, smells and general pressures of the outside world. My brain couldn't cope with it all and I think that is why my memories are a little hazy.

I needed help with everything. Mum had to be my full time carer and I never wanted that for her. We had made the decision whilst I was in rehab to put the house up for sale. Knowing that it was the worst time to try and move, we felt that we needed a bungalow and to be closer to Andrew. Family are the centre of the universe and they are important. We never know when the switch may be flicked to end our lives and then time's up. Life really is too short. So we began to look at new areas and bungalows. It would need to be a complete change of life but, as we had already had an enormous change and we were trying to work out how to manage anyway, it seemed like a sensible time to do it. We weren't in a hurry and our efforts were tentative but the process was underway.

Life at home was difficult for all of us. We had all experienced trauma but even though the trauma had the same cause (i.e. me), our actual personal experiences were very different. Talking about our respective memories was triggering and none of us really knew how best to handle it.

My family experienced the worry and the heartache of being unable to see me. They were denied the opportunity to spend time with me – even when the future looked bleak for me. They were unable to see for themselves that I was being cared for and they couldn't hold my hand or provide the reassurance that I know they would have wanted to give – even though part of me is grateful that they didn't have to endure seeing me like that. They also had the practical side of dealing with my life outside of the hospital, finances, work etc. And then they had decisions to make. Putting the DNR in place must have been devastating and I know they had conversations about what I would want if the worst were to happen – specifically with regard to organ donation. There were some very difficult telephone conversations with the ICU team which I know have stuck in their heads. They then had the relief as I started to respond coupled with the total loss of contact with the medical staff once I was able to make decisions and receive medical information for myself.

My experience, on the other hand, was obviously completely different because I didn't, and still don't, remember the times when it looked like I may not survive. I wasn't aware of any of that. But waking up in delirium with no one familiar and no idea what was going on must have been very frightening and confusing. I say 'must have' because I don't have the memories which makes it even more difficult to deal with. Because what is there to deal with if I can't remember?

I often wonder whether things would have been different if I could have had that physical contact with family and friends. I had a CD of their voices but the physical touch of love cannot be conveyed on a CD. Once I started coming round and was unresponsive for a week I must have been trapped somewhere else and maybe my family and friends could have brought me back quicker. I won't ever know.

The memories I did have were often terrifying and I was still trying to process them by the time I went home. I was still remembering things that either happened in my mind or really happened and often I couldn't distinguish between the real and the imagined. There was no one to ask and so I was emotionally battling to figure everything out.

And then there were physical limitations that we were still adapting to – and still are. We needed to work out how I was going to manage around the house and, after that, we needed to work out how I was going to leave the house. Life with a wheelchair is undoubtably different and there are more considerations than I thought possible – until you are faced with them you have no idea. I didn't.

I had been prone to falls in the hospital and I wasn't totally steady on my feet. But suddenly I was faced with stairs twice a day and the freedom of being at home. I wasn't really well enough to go anywhere near the kitchen. None of that had been covered in rehab and I wasn't ready for it. So I needed constant supervision and care.

I still had pain in my arm from the newly healed ulcer and

nerve pain/numbness in my right arm and left leg. I was on a number of medications and I wasn't able to dispense them myself because, even though I knew what medication I took, my right hand didn't have the dexterity to handle small pills and I dropped them a lot. So much so that Mum took over managing my medication as we were concerned about the dogs and bits of medication on the floor. Plus it was very frustrating.

And then there were financial considerations. I wasn't able to return to my normal job. I was unable to drive (and didn't have a suitable car anyway) and needed to understand the benefits system in order to survive. I'm not very good at asking for help so spent a long time trying to work it all out by myself and it is complicated.

So going home was overwhelming for all of us.

Unfortunately after a day or two I did have my first fall. I was standing from the toilet and fell to the side. I hit my right elbow very hard and I was upset as Mum had to come and help me – luckily I fell back onto the toilet rather than onto the floor. Due to the lack of sensation I didn't realise that my right arm was extremely bruised and I had obviously caused some damage to my elbow (which I could feel just by touching it with my left hand). This fall made us both a bit more nervous!

As I predicted, we were left floundering for quite a while. It was very difficult to see a way forward and we felt like we had very little support. After a few weeks of wondering what on earth we were going to do, I had a call from the community OT who wanted to come out and see me and we made an appointment for her to visit. She came out and looked at my walking as well as aids I had and she ordered more that could help – like a trolley on wheels to help me move food from the kitchen to the living room as I was using two crutches to mobilise at home. She was helpful but only visited me the once.

At the same time I also had a call from the community physio who wanted to come out so she made an appointment and visited. She watched me walking and gave me some exercises to try and build on strength. I was still very weak but I was able to get around at home. Not always safely as I was still tripping over my left foot a lot. The physio came back once more and then followed up with a phone call but, if I'm honest, she didn't really tell me anything that I didn't already know.

I had seen a new rheumatologist, Dr Barclay, whilst I was in ICU and I did have a follow up appointment with him. During my admission I had seen him a few times – he even popped in when I was in A&E after the head injury! During my follow up appointment he was very thorough and we spent a long time with him. I felt reassured that he would coordinate my care but sadly I didn't see him again after this. Instead, I was passed over to his registrar who hadn't even read my notes when I saw him. It was quite disappointing because I felt lost again and my history was so complex that I needed any doctor treating me to understand what had happened. A five minute appointment isn't long enough to convey that volume of information. I hadn't been referred to see any other consultants – despite being in heart failure still and having kidney disease.

Given the lack of trust I had in the registrar, I decided to go back to my private rheumatologist. I had been seeing him prior to catching Covid and he was aware of my more recent history because Dr Barclay had been liaising with him. He was concerned about not being part of a team of doctors caring for me because he felt I needed a team around me but I simply didn't have a team and I was back to having no one coordinating anything. So going back to him managing my lupus at the least was helpful. He knew me and I was comfortable with him. He was calm and patient and he treated me like a human being. I trusted his judgement as he was the doctor who listened and finally diagnosed my lupus after years of symptoms that no one ever put together.

Unfortunately my GP was still not seeing patients and it was very difficult to get in to see her but after a few weeks she did call and ask me to go in so she could see me. Despite the entire country being back to 'normal', they were still only allowing a couple of people into the waiting room and appointments were difficult to get but I went along and she saw me in the downstairs consulting room. Unfortunately it was placed in a really awkward position and the room was very small so once I got the wheelchair in, there was very little room for anything else and it was quite difficult to shut the door. She was very nice and we talked about what had happened and she asked if I needed anything. She referred me for further support at the hospital as the community team were very much short term and couldn't do much for me. But otherwise there wasn't much she could do.

I was referred to physio at the local hospital and she gave me a splint for my left foot. She also referred me to get one properly fitted because the one she had didn't fit completely. Because I had fallen she gave me some balance exercises to try and make things a little safer. She also referred me for hydrotherapy and to a different OT at the hospital called Sarah. I finally felt like someone had noticed I was there.

Sarah was excellent when I got an appointment. She seemed very competent. I desperately needed someone to coordinate my care because nobody was doing it and we felt that I should have been having more follow ups. My experience of a new geographical area later on would highlight this as a number of much needed referrals were finally made, but it took for us to move house to access that level of support. Sarah finally referred me for the wheelchair that rehab hadn't even tried to get for me.

They took measurements and a wheelchair was made for me. I felt very lucky because I know they are quite difficult to get if you have some mobility but I really needed it and the chair they made was finally comfortable – it was also a tank but unfortunately so was I by now thanks to copious amounts of steroids, general medication and reduced mobility!

I started hydrotherapy and loved it. Being in the water was calming and felt easy. The physio there worked with me on strengthening arms and legs and we used the water as resistance. I was also able to have a shower because at home we only had a bath and I was unable to get into it. That first shower was amazing. Until, that is, the shower chair ran away from me and I fell, completely naked, onto the floor. Worst nightmare again! Helpless, on the floor, wet, naked and totally mortified. I also pulled the shower off the wall in the process. People came running – including Mum who was waiting outside and the male staff member who was in the pool area. Mum protected my modesty with towels whilst the physios had to get me up. It was extremely difficult and certainly ruined my first shower since leaving hospital. In every session after that I made sure that the shower chair was firmly wedged against the wall before getting undressed! Shower aside, hydrotherapy helped enormously – even just with general movement and exercise. I was disappointed when it ended after six weeks.

As we were getting ready to move I suddenly got more referrals through – perfect timing! I'd had heart failure so got a call from cardiology. They had requested an ECG which appeared to be normal and an echo to look at my heart. They discovered a flap in my heart which should have closed when I was a baby but hadn't. Apparently 25% of people have this abnormality but it rarely causes problems. One problem that it can cause though is a stroke and, given that I had one in ICU, the cardiologist wanted to do a 'bubble scan' as a precaution to see if there was any communication between the left and right sides of my heart. He didn't think it would be positive but wanted to be thorough. When I later had this scan we found out that I did indeed have communication between the left and right side of my heart and this could have caused the stroke. Sometimes it needs closing but because I'd had a pulmonary embolism before I got Covid, I was already on blood thinners so closure wasn't needed. Which does lead me to question the need for the bubble scan in the first place. It isn't particularly invasive but does involve scanning whilst injecting bubbles into a vein so isn't exactly pleasant and inevitably it causes anxiety.

Unfortunately my kidney function took a bit of a nose dive again once I was home. After having kidney failure and needing dialysis in ICU, I was very lucky that my kidney function recovered back to where it had been before. My kidneys hadn't been functioning as they should for a number of years but they got back to a similar level which was amazing. After I left rehab the function dropped and I was referred to a nephrologist. I saw him just before we moved and he just said it might be lupus … but also might not be lupus. Well I could have worked that much out by myself! He said my kidneys were functioning at around 45% but seemed to be stable at that level so he discharged me and recommended regular bloods just to monitor. They remain stable and no action is required currently but they are monitored regularly.

I am still unable to feel my left foot and ankle or around 85% of my right arm. I am able to use my arm luckily even though it feels strange. I can't feel some of my fingers though and my right hand/arm is quite weak. Not feeling the tip of my index finger affects my fine motor skills as I am not able to pick up small objects. It's very frustrating – particularly because crafting is an important hobby of mine. I have taught myself to write with my left hand and, whilst I am slow and my writing isn't neat, it is legible and can only get stronger.

The lack of feeling in my left foot affects walking as my foot drops and I tend to drag it or trip over it. I also have nerve pain in my lower leg and it becomes quite discoloured. I have more recently had swollen feet and ankles which could be linked to my heart failure, my kidneys, lack of movement or a mix of all three. I also still have a very wobbly left knee which just gives way at any point – thankfully this seems to be the only stroke related deficit I have. I am incredibly lucky for that.

My breathing has improved immeasurably but I still get very short of breath and there is significant scarring left behind. Luckily it isn't thought that this will worsen. My lung function is at 60% – I actually consider myself lucky that it has recovered to that level given the severity of the infection. My lungs were filled with this sticky substance that was never really identified (to our knowledge) and recovery was very uncertain so they have done pretty well.

Then there is long Covid. My breathlessness is partly down to long Covid. I also suffer with extreme fatigue. I have always suffered with this but it is now severe and really affects what I am able to do. I have to sleep most days and find that the simplest activities tire me out very quickly. It isn't like the normal tired of having a bad night. It is crippling tiredness where the only choice is to sleep. It makes it very difficult to do anything because I suffer for it afterwards. To anyone else it may look like I can do things but I will push myself to give that impression. The crash that follows isn't usually visible to anyone except Mum because I purposely don't show it and it can last for days afterwards.

My voice is still breathy and quiet. It is also lower than before and my singing voice is gone – if I try to take it higher than my speaking voice there is nothing there. This is probably the hardest thing as anyone who knows me will understand because singing has always been something I've loved – whether in a choir or by myself. I miss being able to belt out my favourite tunes in the kitchen and being able to sing with my friends. I have also been left with some swallowing problems after the ventilation and tracheostomy. I am hopeful that, like with walking, I just need to learn how to swallow, talk and sing again with speech therapy and have found a speech therapist who has helped me enormously.

As time has passed since my discharge, I have spent a lot of time thinking about what I actually want now. Me. No one else. My hopes and goals weren't always taken into consideration previously and I did what was expected of me but I have to now prioritise for myself. So number 1 is my voice, number 2 is some fine motor skills back, number 3 is independence. The big one. How am I going to live as an independent adult in the future? Getting back to driving is one hurdle. I can't drive Mum's car anymore because of the lack of feeling in my left foot but I can drive an automatic and now have one – with a hoist in the boot for my chair. Independence is finally within my grasp – once I am brave enough!

Managing the wheelchair is another topic entirely. There is no doubt that it will stop me from going out on my own. I have always been very against an electric wheelchair because I really want to keep physically active but I will admit that it is more difficult than I thought it would be and three years later I have just been given a powerchair.

Then there is the emotional side. Unfortunately my ICU experience has left me with some PTSD and I know that I will need to deal with this. There are lasting nightmares which occupy my mind and there are also plenty of real memories which are unpleasant. But those memories will always be there and I am hopeful that I am slowly beginning to accept them. I spend a lot of time lost in thought and maybe this is just how it's going to be for a while. I have had a lot of support from amazing staff in ICU and have been back to visit. I'll admit it was a bit of a shock because for some reason I didn't prepare myself to see real patients on ventilators. I think I imagined they would just take me to an unoccupied space but of course they didn't.

I saw the room I was in when I was the most poorly. I don't remember it at all but it was bigger than I expected and also a little more private with frosted sliding glass doors. There were a lot of machines obviously and a desk for the staff member assigned to that room. Mum recalls this room from when she was called into the hospital at 2am to say goodbye to me as I was taken into intensive care. I was awake and talking a bit but I wasn't really lucid and soon after I was sedated. She remembers this room well. I think I got a sense of just how unwell I was whilst I was in this room. It was touch and go more than once.

I was moved to a separate area when I started to improve slightly, but was still ventilated and we revisited that room. This was more familiar as I know I spent a while watching the lady in the bed opposite and I recognised her cubicle. This was where I firstly didn't regain consciousness for a week after sedation was removed and where I subsequently experienced the delirium.

I will forever be in awe of the lovely Frances who looked after me when I was in ICU and who took me back there to visit then gave me a lot of her time to talk through things that had surfaced. She enabled me to realise that the things I am suffering with are perfectly normal after being so sick. Sadly the lady opposite me in ICU didn't survive which did affect me even though I don't really know her. I do remember her name and I remember hearing it a lot as the nurses were communicating with her.

Everyone who works in ICU is, in my eyes, a superhero because it is an extraordinary environment and must hold a lot of sadness. I watched families visiting their sick relations and some of them were quite distressed. My conscious self now feels very sad that because of the Covid restrictions there was no one to hold my hand whilst I was so incredibly unwell. I had nurses and carers with me all the time but not the touch and sound of family. I think about how different it could have been if they were able to be there. But in the same respect, I am grateful on their behalf that they weren't because I think the memories of that particular time would have been disturbing. Would the comfort of seeing me and knowing exactly what was going on have compensated for the distress of seeing me like that? I think it could have been both a blessing and a curse because not knowing everything would also have been awful.

My experience made me a 'patient'. I have spent so long in hospital and going backwards and forwards to appointments that I honestly don't know what to do when they are finished. Life without the next hospital appointment to go to seems very frightening. I realise that sounds completely insane. Once the appointments and visits stop then it feels like it's all over and I have to actually live my life. Carve out something entirely new. If the appointments have stopped then does that mean this is it now? Recovery is over? Do I now accept that I'm 'finished' and work out how my life will be or am I still meant to be fighting? The problem is, I have spent so long as a patient that I have completely forgotten how to be a person. How to live. How to just be.

But I will. If my story has shown me anything it is that there is always hope. Even when things look bleak. It has also taught me that family and friends are everything and that the ones who count will be there no matter what. I feel incredibly fortunate to have a handful of people who supported not only me but also Mum, through one of the worst experiences of our lives. I will be forever indebted to these people.

Dog walking

ACKNOWLEDGEMENTS

My immediate family and close friends went through hell and I am full of admiration for and gratitude to every one of them. I don't think I could be so strong in the same situation and I know they fought for me, prayed for me and supported me. I can only imagine how difficult it must have been.

A very special thanks has to go to my lovely mum who has contributed to the writing of this book and who has supported me and encouraged me to allow it to reach a wider audience.

I can never thank the ICU team who saved my life enough because without them this book would never have happened. A group of truly wonderful individuals who do an incredibly difficult job. To Francis who is wonderful and who has been my support contact since I left ICU and who has guided and supported my recovery and to all the physios, doctors, nurses and other health professionals who have aided me in my recovery.

A huge thank you also has to go to my friends and family for supporting me to turn what was a therapy process to a project which can hopefully help and encourage others as well as document the effect of Covid on both me and those around me. To my GP Amy and all the team at my GP surgery who have supported my recovery and taken time to listen to my story – I would not be as well as I currently am without their support.

Another thank you to those who read the book in advance of me sending it to my publisher, in particular my neighbour

Brian and my Godson's partner Leah who both provided very valuable feedback and encouraged me to take the book further.

Thank you to my editor Lorna Brookes at Crumps Barn Studio who has given me the courage to bring my book into the wider world. I am so appreciative of her enthusiasm and positivity right from the first time she read the book.

We will certainly never forget Covid but I am hopeful that eventually it will have changed our lives for the better. After all, where there is love there is always a glimmer of hope. And hope combined with love, family and friends underpins absolutely everything.

Am I a Covid victim? No. I choose survival. It may be a different life to the one I had before but I have learnt to treasure family and friends because all it takes is the flick of a switch to turn it all upside down and you can never get wasted time back.

If you loved this book,
please leave a review ...

The best place to leave a review is on Waterstones.com or on Amazon

Reviews make our authors happy and help us to share news of their work with more readers. Thank you for your support!

Did you know that this book has been ethically printed in the UK on FSC certified paper made from a mixture of fibre from responsibly managed woodlands and recycled material?

Crumps Barn Studio
crumpsbarnstudio.co.uk

ABOUT THE AUTHOR

Jen has worked in early years for a number of years and has watched many children grow into successful and happy adults. Her work is her passion and she is hugely proud of all the children she has cared for over the last 30 years.

Jen loves anything crafty and can often be found with a paint brush, sewing needle, knitting needles or anything where she can create something.

She lives at home with her mother and two dogs, Poppy and Ivy, in a small village in Wiltshire.

Discover your next heartfelt read ...

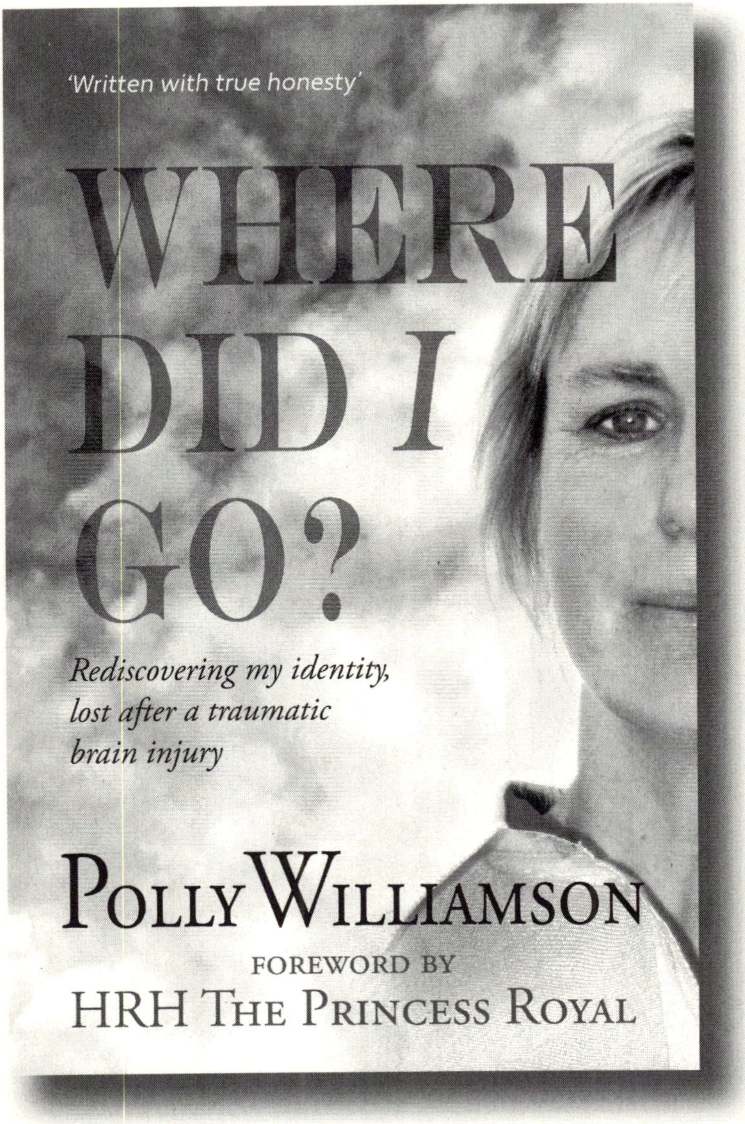